Mastering Mechanical Desktop: Native Solids

RON K. C. CHENG
The Hong Kong Polytechnic University

 Autodesk.
Press

I(T)P An International Thomson Publishing Company

Albany • Bonn • Boston • Cincinnati • Detroit • London • Madrid
Melbourne • Mexico City • New York • Pacific Grove • Paris • San Francisco
Singapore • Tokyo • Toronto • Washington

Trademarks

AutoCAD, AutoCAD Designer, Mechanical Desktop, AME, AutoSurf, and NURBS are a registered trademarks of Autodesk, Inc.
The ITP logo is a trademark under license.
Windows is a registered trademark of Microsoft Corporation.

Copyright ©1998
PWS Publishing Company
Autodesk Press imprint
an International Thomson Publishing Company

Printed and bound in the United States of America.
1 2 3 4 5 6 7 8 9 10 — 01 00 99 98 97

For more information contact:

Autodesk Press
3 Columbia Circle, Box 15-015
Albany, New York 12212-5015

International Thomson Publishing Europe
Berkshire House
168–173 High Holborn
London WC1V 7AA
England

Thomas Nelson Australia
102 Dodds Street
South Melbourne, 3205
Victoria, Australia

Nelson Canada
1120 Birchmont Road
Scarborough, Ontario
Canada M1K 5G4

International Thomson Publishing Southern Africa
Building 18, Constantia Park
240 Old Pretoria Road
P.O. Box 2459
Halfway House, 1685 South Africa

International Thomson Editores
Campos Eliseos 385, Piso 7
Col. Polanco
11560 Mexico D.F., Mexico

International Thomson Publishing GmbH
Königswinterer Strasse 418
53227 Bonn, Germany

International Thomson Publishing France
Tour Maine-Montparnasse
33, Avenue du Maine
75755 Paris Cedex 15, France

International Thomson Publishing Asia
221 Henderson Road
#05-10 Henderson Building
Singapore 0315

International Thomson Publishing Japan
Hirakawacho Kyowa Building, 31
2-2-1 Hirakawacho
Chiyoda-ku, Tokyo 102
Japan

Assistant Editor: Suzanne Jeans
Production Editor: Andrea Goldman
Manufacturing Buyer: Andrew Christensen
Marketing Manager: Nathan Wilbur
Interior Design/Cover Image: Ron C. K. Cheng
Cover Design: Autodesk, Inc.
Cover Printer: Phoenix Color Corp.
Printer & Binder: Courier–Westford

Library of Congress Cataloging-in-Publication Data

Cheng, Ron.
    Mastering Mechanical Desktop: native solids / Ron K. C. Cheng.
        p.   cm.
    Includes index.
    ISBN 0–534–95108–2
    1. Engineering graphics.   2. Autodesk Mechanical desktop.   3. Engineering design--Data processing.
4. AutoCAD (Computer file).   I. Title.
T353.C5178   1997   97–30211
620'.0042'02855369--dc21   CIP

Mastering
Mechanical Desktop:
Native Solids

Contents

Preface

AutoCAD Release 13 solids, referred to as "native solids" in AutoDesk, are 3D solid objects that are created by constructive solid geometry techniques. The ancestor of the native solids is the Advanced Modeling Extension (AME) solids of AutoCAD Release 12/Release 11.

This book is intended for those who have a general understanding of how to use AutoCAD as a computer-aided design tool and want to learn how to create 3D solid objects with AutoCAD.

The first chapter of this book begins with an overview of solid objects and delineates how solid models can be built in a computer. You will also learn how a native solid is compatible with Release 12 AME solids, and Mechanical Desktop objects.

Chapter 2 guides you to create various primitive solids and to create a complex solid by using two Boolean operations, union and subtract. In Chapter 3, you will extrude and revolve 2D regions and polylines into solids, and learn how to apply the Boolean operation intersection in model making. Then, in Chapter 4, you will work on a complex thin shell solid with internal bosses and webs. You will also learn how to use the utility commands on a solid.

Outputting a 3D solid model in 2D paper format for documentation is sometimes necessary. Chapter 5 will tell you how to obtain orthographic and 3D views in a paper document.

To further enhance your knowledge of solid modeling, Chapter 6 guides you to produce a 3D structural framework. Finally, Chapter 7 explores the use of constructive solid geometry technique on an architectural project. The end of this book contains a brief explanation of all the solid modeling commands that were used. After working through all the projects in this book, you should have a good understanding of solid modeling techniques and be able to apply the techniques to projects of your own.

Acknowledgments

This book never would have been realized without the contributions of many individuals.

I am grateful to the following reviewers for their thoughtful suggestions and help:

- Robert A. Chin, Department of Industrial Technology, East Carolina University
- Hollis Driskell, Department of Drafting and Design, Trinity Valley Community College
- Michael Stewart, Department of Engineering Technology, University of Arkansas
- Ed Wheeler, Engineering Department, University of Tennessee at Martin

Several people at PWS Publishing also deserve special mention, particularly Bill Barter, Jonathan Plant, Suzanne Jeans, Andrea Goldman, Tricia Kelly, and Monica Block.

Chapter 1

Introduction

There are three kinds of 3D models in a computer — the wireframe model, the surface model, and the solid model. A 3D-wireframe model is the most primitive type of 3D object. It is a set of unassociated line and arc segments that are put together in a 3D space. The line and arc segments serve only to give the pattern of a 3D object. There is no relationship between them. As such, the model does not have any surface information or volume information. It has only data that describe the edge of the 3D object. Because of the limited information provided by the model, the use of wireframe models is very confined.

The second type of 3D model, the surface model, is a set of surfaces that are put together in a 3D space to give the figure of a 3D object. When compared to a 3D-wireframe model, a surface model has, in addition to edge data, information on the contour and silhouette of the surfaces. You can use the surface models in a computerized manufacturing system or to generate photo-realistic rendering or animation.

In regard to information, a 3D-solid model is superior to the other two models because a solid model in a computer is an integrated mathematical data that contains information not only on the surfaces and edges, but also on the volume of the object that the model describes. In addition to visualization and manufacturing, you can use the data of a solid model for design calculation.

1.1   Solid Modeling Applications

There are many ways to represent a solid in a computer. The following briefly describe some of the methods.

Pure Primitive Instancing Method

This method predefines a limited range of solid objects. To make a solid, the user supplies the values of the parameters to the system.

Generalized Sweep Method

This method creates solid objects by sweeping a 2D or 3D lamina along a 3D curve.

Spatial Occupancy Enumeration Method

This method divides the entire 3D space into a number of cubical cells. Solids are represented by listing the cells that the solids occupy in the 3D space. Accuracy of the solid being represented is a function of the size of the cubical cells. However, a smaller cubical size increases the file size tremendously.

Cellular Decomposition Method

This method is similar to the spatial occupancy enumeration method in that the solids are represented by listing the cells in the 3D space. However, the cells are not necessarily identical or cubical in shape. As a result, this method requires less memory.

Constructive Solid Geometry Method

This method provides a range of primitive solids in a way that is similar to the pure primitive instancing method mentioned above. In addition, it provides an extra facility for the user to perform Boolean operations on the primitive solids to produce a complex solid.

1.2   AutoCAD Solid Modeling

In the previous releases of AutoCAD, the solid modeling tool was a separate package, known as the Advanced Modeling Extension (AME). In Release 13, the package is an integral part of the AutoCAD basic package.

The set of solid modeling commands provided by the AutoCAD basic package uses constructive solid geometry techniques. Constructive solid geometry is a building block approach to create 3D solid models. The system provides tools to create solids of simple geometric shape as building blocks, and Boolean operations to combine the solids together. Any complex solid model is created by combining simpler solids together. The simplest solids are called primitives. The six primitive types available are box, sphere, cylinder, cone, wedge, and torus. See Figure 1.1.

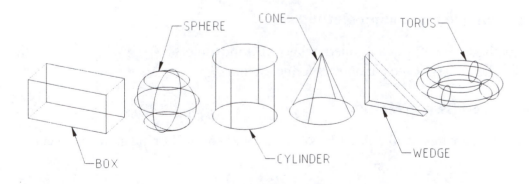

Figure 1.1 Primitive solids

The repertoire of complex solids that can be composed from these six types of primitive solids is very limited. To extend the scope of model creation, AutoCAD allows you to make solid objects by extruding a 2D object along a line and revolving a 2D object about an axis. The 2D object can be either a region or a closed polyline. Figure 1.2 shows an extruded solid and a revolved solid.

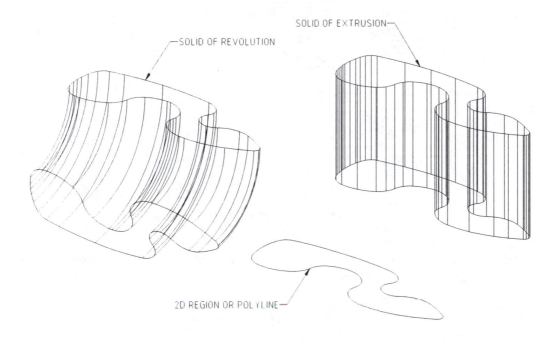

Figure 1.2 Extruded solid and the revolved solid

To compose two solid objects together to form a complex solid, you can use one of the three types of Boolean operations: union, subtraction, and intersection. A union of two solids creates a solid that encloses all the volume enclosed by the first and the second solid. A subtraction of two solids creates a solid that encloses the volume of the first solid but not the second solid. An intersection of two solids creates a solid that encloses the volume of the first solid that is also contained in the second solid. Figure 1.3 illustrates the effect of these Boolean operations on a solid box and a solid cone.

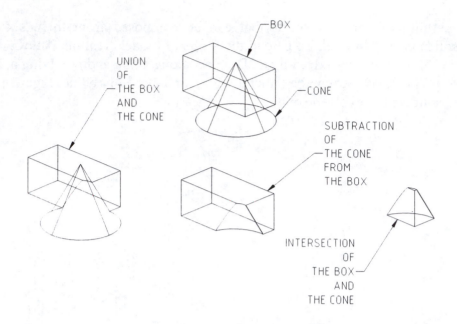

Figure 1.3 Boolean operations

As a further refinement to a solid, you can chamfer or fillet the edge. Figure 1.4 shows a solid box, a chamfered box, and a filleted box.

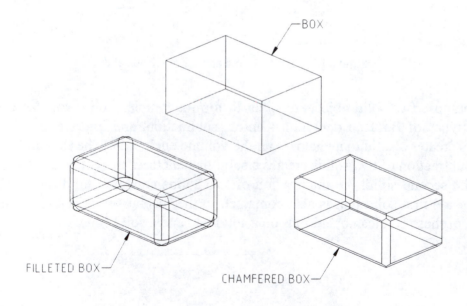

Figure 1.4 Solid box, filleted box and chamfered box

1.3   Compatibility

AutoCAD is forward compatible. A solid that is created with AME of Release 12 can be upgraded to a Release 13 native solid by using the AMECONVERT command.

To convert an AME solid, click the AME Convert item from the Solids toolbar to run the AMECONVERT command.

[Solids]   **[AME Convert]**

Command: **AMECONVERT**
Select objects: **[Select a R12 AME solid.]**
Select objects: **[Enter]**

You might have learned that the SAVEASR12 command can save a Release 13 AutoCAD drawing as a Release 12 format. However, you must not use this command to save a Release 13 solid to Release 12 version. Because the data structure of Release 13 is entirely different from that of Release 12, saving a Release 13 solid to Release 12 format can result in a lost of solid data.

An AutoCAD native solid is 100% compatible with objects created by Mechanical Desktop. Mechanical Desktop of AutoCAD contains two components: AutoSurf and AutoCAD Designer. AutoSurf is an application for creating 3D NURBS surface models. AutoCAD Designer is an application for creating 3D parametric solid models.

You can use a 3D NURBS surface to cut a native solid, as well as to convert a native solid into a set of 3D NURBS surfaces. The commands which are available in AutoSurf are AMSOLCUT and AM2SF.

If you have installed AutoSurf properly and have included the Surfaces item in the pull-down menu, you can select the Edit Solid item from this pull-down menu to run the AMSOLCUT command. See Figure 1.5.

<Surfaces>   **<Edit Solid>**

Command: **AMSOLCUT**
Select solid to cut: **[Select a native solid.]**
Select surface: **[Select a NURBS surface.]**
Portion to remove: Flip/<Accept>: **[Enter, if the direction is correct.]**

> **Note:**
> In the delineation that follows, <AAA> <BBB> indicates that you should use the <AAA> pull-down menu and click the <BBB> item from the menu. [CCC] [DDD] indicates you should use the [CCC] toolbar and click the [DDD] item from the toolbar.

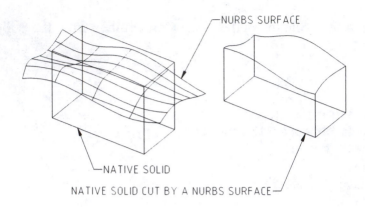

Figure 1.5 Native solid cut by a NURBS surface

To change a native solid into a set of AutoSurf surfaces for further NURBS surface creation, such as making variable fillets, free-form surfaces and other derived surfaces, you can use the Surfaces pull-down menu, and then click the items Create Surface and From ACAD.

<Surfaces>   <Create Surface>   <From ACAD>

Command: **AM2SF**
Face/<objects>: **[Enter]**
Select objects: **[Select a native solid.]**
Select objects: **[Enter]**

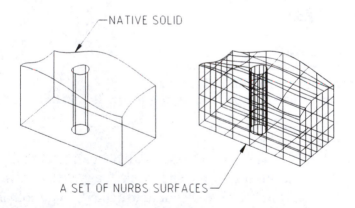

Figure 1.6 Native solid converted to a set of NURBS surfaces

AutoCAD Designer creates 3D parametric solid models. You can convert a native solid to a base solid of AutoCAD Designer, and convert a parametric solid to a static native solid. To convert an AutoCAD Designer parametric solid part to a native solid, you can use the EXPLODE command.

[Modify]   [Explode]

Command: **EXPLODE**

To use a native solid as a base solid feature for further parametric solid feature creation in AutoCAD Designer, you can use the AMNEWPART command. This command is part of the AutoCAD Designer package.

<Parts>   **<Part>**   **<New>**

Command: **AMNEWPART**
Select native solid or (RETURN): **[Select a native solid.]**
....
New part created.

1.4   Isolines and Silhouette Display

As you can see in Figure 1.1, the display of the solids with curved surfaces do not have silhouettes. Instead, you see a series of lines on the surfaces. These lines are called isolines. To control the display of these isolines and silhouette edges, you can manipulate two system variables: ISOLINES and DISPSILH. The ISOLINES variable controls the density of the isolines. If you set it to zero, there will be no isolines. If you set it to a higher value, there will be more isolines. The DISPSILH variable has two settings. If it is one, the silhouette edges appear. If it is zero, the silhouette edges do not show. Silhouettes provide a better picture of the 3D object. However, regeneration of the display will take a long time.

1.5   Summary

In a computer, the solid model of a 3D object provides unique data that contains information on the edges, surfaces, and volume that the model describes. AutoCAD uses constructive solid geometry techniques. You can compose a complex solid from the primitive solids and from the solids of extrusion and revolving. The method of composition can be to unite two solids, to subtract one solid from another, or to form an intersection of two solids. In addition, you can also fillet or chamfer the edges of a solid.

AutoCAD solids are compatible with NURBS surfaces created by AutoSurf and parametric solids created by AutoCAD Designer. A NURBS surface can be used to cut a native solid, and a native solid can be converted into a set of NURBS surfaces. In addition, a native solid can be converted to a new part in AutoCAD Designer, and a parametric solid can be converted to a static native solid model.

Chapter 2

Building Block Principle

Creating a 3D solid model in a computer is very much like manually making a model with building blocks. You can build up a complex object by putting together simpler objects. However, if you compare the two processes, you will find that the former is far more flexible because you can perform subtraction and intersection in addition to union.

To build a complex solid using constructive solid geometry, you need to complete three basic steps. First, you should analyze the complex solid to determine what primitive solids are required and how these primitives are combined. Second, you have to build the primitive solids. Finally, you need to perform Boolean operations on the primitive solids to combine them into one complex solid.

In this chapter, you will perform union and subtraction on a number of primitive solids to create a complex solid. Then, you will refine the complex solid model by filleting and chamfering some edges. Finally, you will slice the complex model into two pieces.

2.1   Analysis

Figure 2.1 shows the rendered complex solid model that you are going to create.

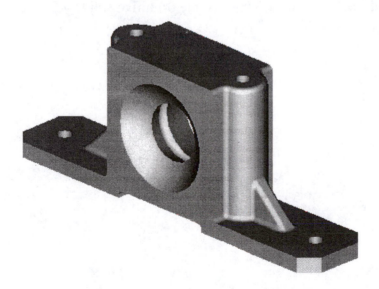

Figure 2.1 Completed solid model

9

To appreciate the building block principle of constructive solid geometry, examine this complex solid model critically. It is symmetrical about its central axis. If you remove the fillet edges and the chamfer edges, you will find that its outer skin consists of three boxes, two cylinders, and two wedges, along with cylindrical, conical, toroidal, and spherical openings. See Figure 2.2.

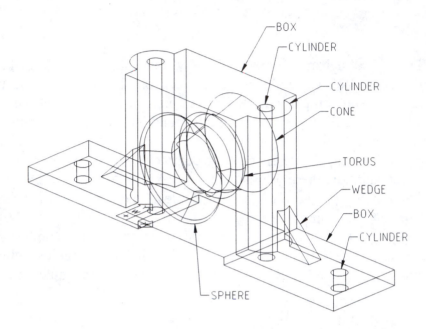

Figure 2.2 Composition of the complex solid

Figure 2.3 shows an exploded view of the primitive solids you need to create in order to obtain the complex solid.

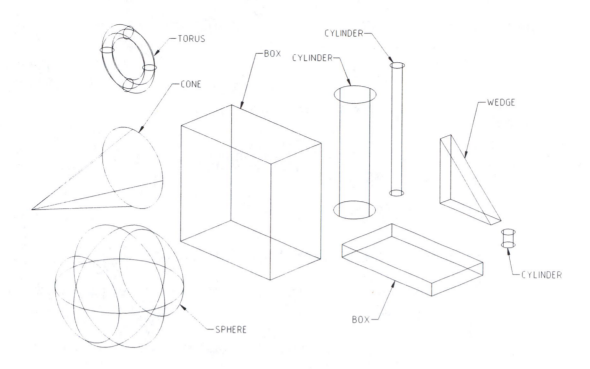

Figure 2.3 Primitives exploded

2.2   Drawing Preparation

Start a new drawing by using the NEW command. Use ACADISO.DWG as the prototype drawing.

    <File>   **<New...>**

    Command: **NEW**

Make a new layer called SOLID. Set its color to yellow. You will create the solid model on this layer.

    <Data>   **<Layers...>**

    Command: **DDLMODES**

Layer	Color
SOLID	**yellow**

Current layer: **SOLID**

When you are working in 3D space, you should set the UCS (User Coordinate System) icon to sit on the origin position. Run the UCSICON command. Preferably, you should set the display to an isometric view, use the VPOINT command to set the viewing direction to rotate 315° in the XY plane, and 25° from the XY plane.

<Options>   **<UCS>**   **<Icon Origin>**

Command: **UCSICON**
ON/OFF/All/Noorigin/ORigin : **OR**

<View>   **<3D Viewpoint>**   **<Vector>**

Command: **VPOINT**
Rotate/<View point> : **R**
Enter angle in XY plane from X axis : **315**
Enter angle from XY plane : **25**
Regenerating drawing.

2.3 Primitives Building, Subtraction, and Union

You will use AutoCAD solid creation commands to make the required primitives, and use AutoCAD Boolean operation commands to integrate them into a single complex solid.

Run the BOX command to build a solid box. Then run the ZOOM command twice to get a view similar to Figure 2.4.

[Solids]   **[Box]**   **[Corner]**

Command: **BOX**
Center/<Corner of box> : **0,0,0**
Cube/Length/<other corner>: **@90,50,90**

[Zoom]   **[Zoom Extents]**

Command: **ZOOM**
All/Center/Dynamic/Extents/Left/Previous/Vmax/Window/<Scale(X/XP)>: **E**
Regenerating drawing.

[Zoom]   **[Zoom Scale]**

Command: **ZOOM**
All/Center/Dynamic/Extents/Left/Previous/Vmax/Window/<Scale(X/XP)>: **.7X**

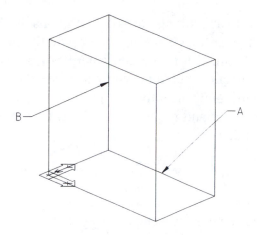

Figure 2.4 Solid box created

Use the UCS command to set the UCS to a new position. Then, create a solid cone by using the CONE command. See Figure 2.5.

[UCS]   **[Z Axis Vector UCS]**

Command: **UCS**
Origin/ZAxis/3point/OBject/View/X/Y/Z/Prev/Restore/Save/Del/?/<World>: **ZA**
Origin point : **.X** of **MID** of **[Select A (Figure 2.4).]**
(need YZ): **MID** of **[Select B (Figure 2.4).]**
Point on positive direction of Z-axis: **@0,-1**

[Solids]   **[Cone]**   **[Center]**

Command: **CONE**
Elliptical/<center point> : **0,0,0**
Diameter/<Radius>: **30**
Apex/<Height>: **100**

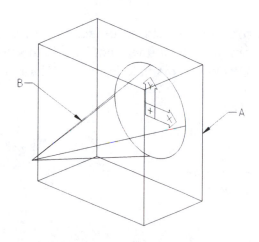

Figure 2.5 Solid cone created

Figure 2.5 shows a solid cone sitting partially within a solid box. In reality, this is not possible. You cannot put a solid object partly or wholly inside another solid object without damaging one of them. But in the computer, they are simply two unrelated data that represent two solid objects. Therefore, you can place them anywhere you like. You can manipulate each of them independently.

Run the SUBTRACT command to perform a Boolean operation to subtract the cone from the box. See Figure 2.6.

[Modify]   **[Subtract]**

Command: **SUBTRACT**
Select solids and regions to subtract from...
Select objects: **[Select A (Figure 2.5).]**
Select objects: Select solids and regions to subtract...
Select objects: **[Select B (Figure 2.5).]**
Select objects: **[Enter]**

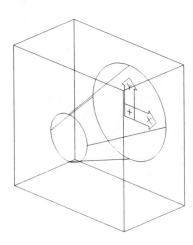

Figure 2.6 Solid cone subtracted from the solid box

After subtraction, the cone and the box become a single complex solid — a solid box with a conical opening. From now on, you will not be able to separate the primitives.

To continue, use the TORUS command to build a solid torus. See Figure 2.7.

[Solids]   **[Torus]**

Command: **TORUS**
Center of torus : **0,0,25**
Diameter/<Radius> of torus: **21**
Diameter/<Radius> of tube: **5**

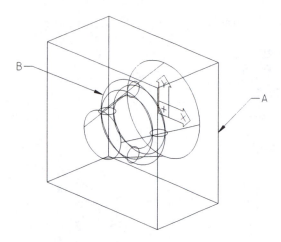

Figure 2.7 Solid torus created

The torus is used to cut a groove. Run the SUBTRACT command to subtract it from the complex solid. See Figure 2.8. Again, you will not be able to separate the torus from the complex solid after the Boolean operation.

[Modify]   **[Subtract]**

Command: **SUBTRACT**
Select solids and regions to subtract from...
Select objects: **[Select A (Figure 2.7).]**
Select objects: Select solids and regions to subtract...
Select objects: **[Select B (Figure 2.7).]**
Select objects: **[Enter]**

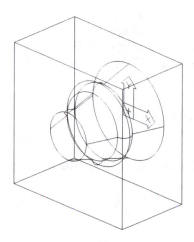

Figure 2.8 Solid torus subtracted from the complex solid

To prepare a spherical recess on the complex solid, create a solid sphere by using the SPHERE command. See Figure 2.9.

[Solids]   **[Sphere]**

Command: **SPHERE**
Center of sphere : **0,0,80**
Diameter/<Radius> of sphere: **45**

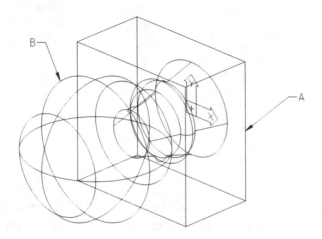

Figure 2.9 Solid sphere created

To cut a spherical recess, run the SUBTRACT command to subtract the solid sphere from the complex solid. After that, set the UCS back to WORLD by using the UCS command. See Figure 2.10.

[Modify]   **[Subtract]**

Command: **SUBTRACT**
Select solids and regions to subtract from...
Select objects: **[Select A (Figure 2.9).]**
Select objects: Select solids and regions to subtract...
Select objects: **[Select B (Figure 2.9).]**
Select objects: **[Enter]**

[UCS]   **[World UCS]**

Command: **UCS**
Origin/ZAxis/3point/OBject/View/X/Y/Z/Prev/Restore/Save/Del/?/<World>: **WORLD**

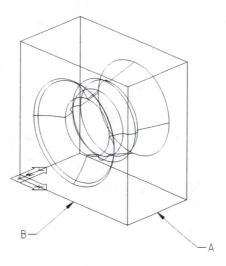

Figure 2.10 Solid sphere subtracted from the complex solid

Now, you have a solid cone, a solid torus, and a solid sphere subtracted from the solid box. To continue, use the CYLINDER command to create a solid cylinder, and run the MIRROR3D command to mirror this solid cylinder. See Figure 2.11.

[Solids]   **[Cylinder]**   **[Center]**

Command: **CYLINDER**
Elliptical/<center point> : **MID** of **[Select A (Figure 2.10).]**
Diameter/<Radius>: **15**
Center of other end/<Height>: **90**

[Copy]   **[3D Mirror]**

Command: **MIRROR3D**
Select objects: **LAST**
Select objects: **[Enter]**
Plane by Object/Last/Zaxis/View/XY/YZ/ZX/<3points>: **YZ**
Point on YZ plane : **MID** of **[Select B (Figure 2.10).]**
Delete old objects? **N**

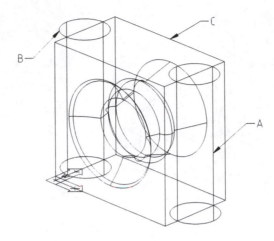

Figure 2.11 Two solid cylinders created

You have practiced using the SUBTRACT command to do subtraction. To unite two or more solids, you can use the UNION command.

Run the UNION command to combine the two solid cylinders with the complex solid. See Figure 2.12.

[Modify]   **[Union]**

Command: **UNION**
Select objects: **[Select A, B, and C (Figure 2.11).]**
Select objects: **[Enter]**

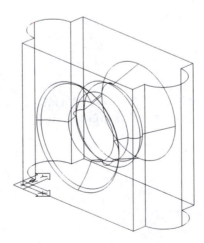

Figure 2.12 Two solid cylinders united with the complex solid

Make another solid box by using the BOX command. See Figure 2.13.

[Solids]   **[Box]**   **[Corner]**

Command: **BOX**

Center/<Corner of box> : **[Select any point on the screen.]**
Cube/Length/<other corner>: **@90,50,10**

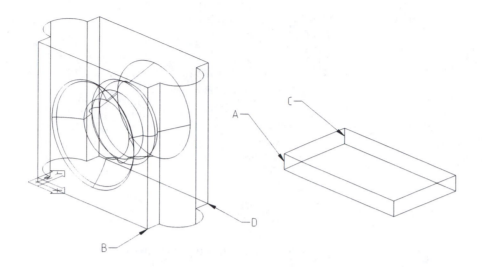

Figure 2.13 Solid box created

The newly-created solid box is a part of the complex solid. To continue, run the ALIGN command to align the box with the complex solid.

Use the middle point of the vertical edge of the box as the first source point, and select a point offset from the lower-right corner of the complex solid as the first destination point. Similarly, set the middle point of another vertical edge of the box to align with a point offset from the other corner of the complex solid. Because the orientation of the objects are correct, you do not need to specify the third source point. Finally, you should specify a 3D transformation for the box to align properly. Otherwise, the system will ignore the Z axis displacement. See Figure 2.14.

[Rotate]   **[Align]**

Command: **ALIGN**
Select objects: **LAST**
Select objects: **[Enter]**
1st source point: **MID** of **[Select A (Figure 2.13).]**
1st destination point: **FROM**
Base point: **END** of **[Select B (Figure 2.13).]**
<Offset>: **@10<180**
2nd source point: **MID** of **[Select C (Figure 2.13).]**
2nd destination point: **FROM**
Base point: **END** of **[Select D (Figure 2.13).]**
<Offset>: **@10<180**
3rd source point: **[Enter]**
<2d> or 3d transformation: **3**

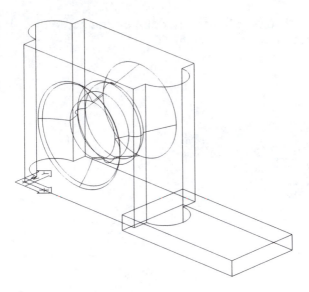

Figure 2.14 Solid box aligned with the complex solid

The complex solid has two wedges. Create a solid wedge by using the WEDGE command. See Figure 2.15.

[Solids]   **[Wedge]**   **[Corner]**

Command: **WEDGE**
Center/<Corner of wedge> : **[Select any point on the screen.]**
Cube/Length/<other corner>: **@50,10**
Height: **50**

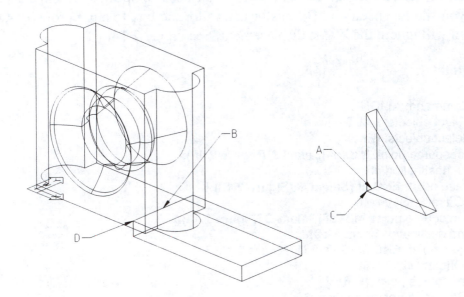

Figure 2.15 Solid wedge created

To align the solid wedge with the solid box, use the ALIGN command. This time, set the middle point of the lower edge of the wedge as the first source point, and set the middle point of one edge of the solid box as the first destination point. Because the width of the solid wedge and the solid box are different, the second source point and destination point will indicate the alignment direction. Also, make a null respond to the third source point and take a 3D transformation. See Figure 2.16.

[Rotate]   **[Align]**

Command: **ALIGN**
Select objects: **LAST**
Select objects: **[Enter]**
1st source point: **MID** of **[Select A (Figure 2.15).]**
1st destination point: **MID** of **[Select B (Figure 2.15).]**
2nd source point: **END** of **[Select C (Figure 2.15).]**
2nd destination point: **END** of **[Select D (Figure 2.15).]**
3rd source point: **[Enter]**
<2d> or 3d transformation: **3**

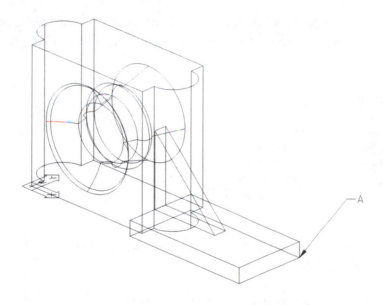

Figure 2.16 Solid wedge aligned with solid box

You will need a hole in the solid box. For this purpose, create a solid cylinder by using the CYLINDER command. Use the FROM object snap and select the lower-left corner of the solid box as the Base point. See Figure 2.17.

[Solids]   **[Cylinder]**   **[Center]**

Command: **CYLINDER**
Elliptical/<center point> : **FROM**
Base point: **END** of **[Select A (Figure 2.16).]**
<Offset>: **@-20,-25**
Diameter/<Radius>: **5**
Center of other end/<Height>: **10**

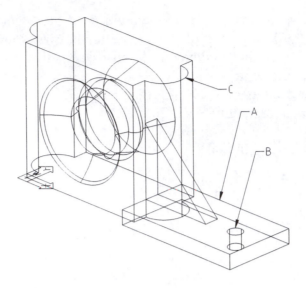

Figure 2.17 Solid cylinder created

The solid cylinder you just created is used to cut a hole in the solid box. Run the SUBTRACT command. There will not be any noticeable visual change to your drawing after the subtraction.

[Modify]   **[Subtract]**

Command: **SUBTRACT**
Select solids and regions to subtract from...
Select objects: **[Select A (Figure 2.17).]**
Select objects: Select solids and regions to subtract...
Select objects: **[Select B (Figure 2.17).]**
Select objects: **[Enter]**

Make another solid cylinder for cutting a hole in the complex solid. Run the CYLINDER command. See Figure 2.18.

[Solids]   **[Cylinder]**   **[Center]**

Command: **CYLINDER**
Elliptical/<center point> : **CEN** of **[Select C (Figure 2.17).]**
Diameter/<Radius>: **5**
Center of other end/<Height>: **-100**

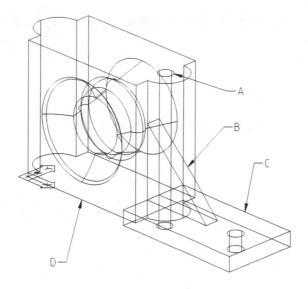

Figure 2.18 Another solid cylinder created

To reiterate, the complex solid is symmetrical about its axis. Use the MIRROR3D command to mirror the solid cylinder, the solid box with a hole, and the solid wedge about the middle point of the complex solid. See Figure 2.19.

[Copy]   **[3D Mirror]**

Command: **MIRROR3D**
Select objects: **[Select A, B, and C (Figure 2.18).]**
Select objects: **[Enter]**
Plane by Object/Last/Zaxis/View/XY/YZ/ZX/<3points>: **YZ**
Point on YZ plane : **MID** of **[Select D (Figure 2.18).]**
Delete old objects? **N**

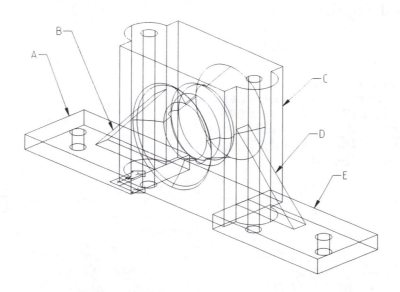

Figure 2.19 Three solid objects mirrored

Run the UNION command to unite together all the objects, except the two solid cylinders. See Figure 2.20.

[Modify]   **[Union]**

Command: **UNION**
Select objects: **[Select A, B, C, D, and E (Figure 2.19).]**
Select objects: **[Enter]**

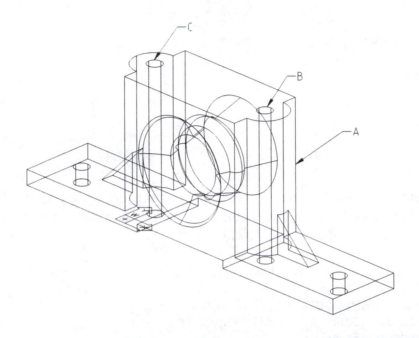

Figure 2.20 All solids, except two solid cylinders, united

To cut two holes, run the SUBTRACT command to subtract the two cylinders from the complex solid. See Figure 2.21.

[Modify]   **[Subtract]**

Command: **SUBTRACT**
Select solids and regions to subtract from...
Select objects: **[Select A (Figure 2.20).]**
Select objects: Select solids and regions to subtract...
Select objects: **[Select B and C (Figure 2.20).]**
Select objects: **[Enter]**

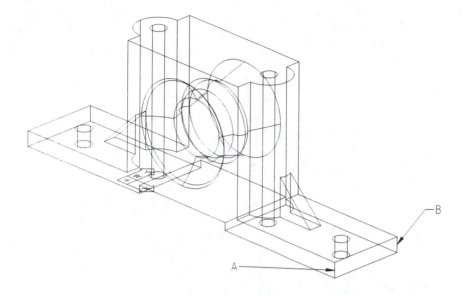

Figure 2.21 Two solid cylinders subtracted from the complex solid

2.4   Chamfering and Filleting

Up to this point, the basic shape of the complex solid is completed. To continue, you will add chamfers and fillets to it.

Apply the CHAMFER command to cut two chamfered edges on the complex solid. See Figure 2.22.

[Feature]   [Chamfer]

Command: **CHAMFER**
Polyline/Distance/Angle/Trim/Method/<Select first line>: **[Select A (Figure 2.21).]**
Select base surface:
Next/<OK>: **[Enter, if the face AB of Figure 2.21 is highlighted. Otherwise, Next.]**
Enter base surface distance : **10**
Enter other surface distance : **10**
Loop/<Select edge>: **[Select A (Figure 2.21).]**
Loop/<Select edge>: **[Select B (Figure 2.21).]**
Loop/<Select edge>: **[Enter]**

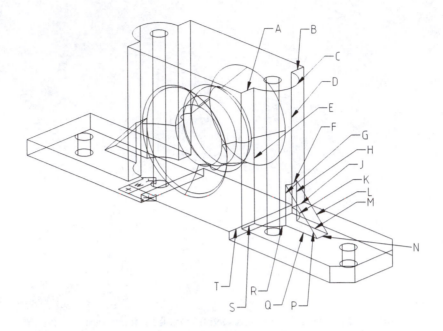

Figure 2.22 Two edges chamfered

After cutting the chamfer, use the FILLET command to fillet 18 edges in one operation. See Figure 2.23.

[Feature]   **[Fillet]**

Command: **FILLET**
Polyline/Radius/Trim/<Select first object>: **[Select A (Figure 2.22).]**
Enter radius: **3**
Chain/Radius/<Select edge>: **[Select B, C, D, E, F, G, H, J, K, L, M, N, P, Q, R, S, and T (Figure 2.22).]**
Chain/Radius/<Select edge>: **[Enter]**
18 edges selected for fillet.

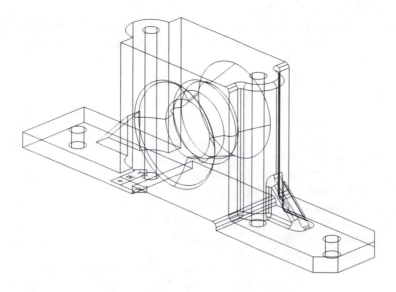

Figure 2.23 Eighteen edges filleted

One side of the model is complete. To continue working on the other side, set the display to a new viewing position by using the VPOINT command. See Figure 2.24.

<View>   **<3D Viewpoint>**   **<Vector>**

Command: **VPOINT**
Rotate/<View point> : **R**
Enter angle in XY plane from X axis : **135**
Enter angle from XY plane : **25**
Regenerating drawing.

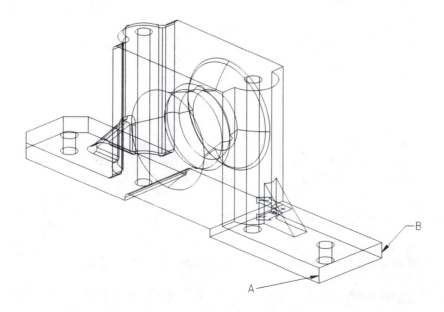

Figure 2.24 New viewing position set

Because the complex solid is symmetrical about its axis, the treatment of both sides are the same. Use the CHAMFER command to cut two chamfered edges on this side of the solid. See Figure 2.25.

[Feature]   [Chamfer]

Command: **CHAMFER**
Polyline/Distance/Angle/Trim/Method/<Select first line>: **[Select A (Figure 2.24).]**
Select base surface:
Next/<OK>: **[Enter, if the face AB of Figure 2.24 is highlighted. Otherwise, Next.]**
Enter base surface distance : **10**
Enter other surface distance : **10**
Loop/<Select edge>: **[Select A (Figure 2.24).]**
Loop/<Select edge>: **[Select B (Figure 2.24).]**
Loop/<Select edge>: **[Enter]**

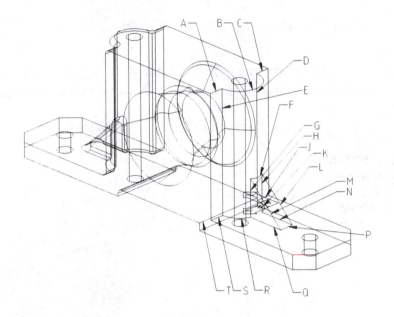

Figure 2.25 Two edges chamfered

The final step to model this complex solid is to cut fillets on 18 edges similar to the other side. Run the FILLET command to fillet all the specified edges in one operation. Finally, zoom to the previous view. See Figure 2.26.

[Feature]   [Fillet]

Command: **FILLET**
Polyline/Radius/Trim/<Select first object>: **[Select A (Figure 2.25).]**
Enter radius: **3**
Chain/Radius/<Select edge>: **[Select B, C, D, E, F, G, H, J, K, L, M, N, P, Q, R, S, and T (Figure 2.25).]**
Chain/Radius/<Select edge>: **[Enter]**
18 edges selected for fillet.

[Standard Toolbar]   **[Zoom Previous]**

Command: **ZOOM**
All/Center/Dynamic/Extents/Left/Previous/Vmax/Window/<Scale(X/XP)>: **P**
Regenerating drawing.

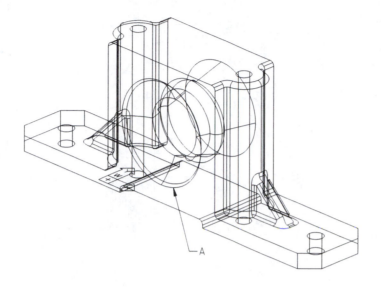

Figure 2.26 Completed complex solid model

The model is complete. Save your file.

<File>   **<Save...>**

Command: **QSAVE**

Before proceeding, you will save another copy of this model to use in Chapter 3. Run the SAVEAS command.

<File>   **<Save As...>**

Command: **SAVEAS**

When the dialog box appears, specify a valid filename.
Now, you should have two identical files.

2.5   Slicing

To further manipulate the complex solid, you will slice it into two pieces. Run the SLICE command and select the BOTH option. See Figure 2.27.

[Solids]   **[Slice]**

Command: **SLICE**

Select objects: **[Select A (Figure 2.26).]**
Select objects: **[Enter]**
Slicing plane by Object/Zaxis/View/XY/YZ/ZX/<3points>: **XY**
Point on XY plane <0,0,0>: **CEN** of **[Select A (Figure 2.26).]**
Both sides/<Point on desired side of the plane>: **B**

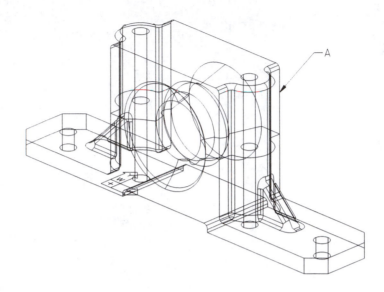

Figure 2.27 Complex solid sliced into two pieces

The model is now sliced into two pieces. To see the effect of slicing, move the pieces apart. See Figure 2.28.

[Modify]   [Move]

Command: **MOVE**
Select objects: **[Select A (Figure 2.27).]**
Select objects: **[Enter]**
Base point or displacement: **45,45**
Second point of displacement: **[Enter]**

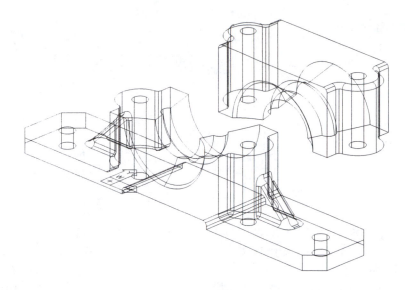

Figure 2.28 Upper portion of the sliced solid moved apart

To see the rendered drawing of the complete model, run the RENDER command. See Figure 2.28.

[Render]   **[Render]**

Command: **RENDER**

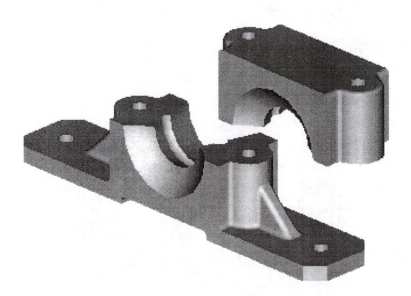

Figure 2.29 Rendered drawing of the sliced solids

The sliced model is complete. Run the QSAVE command.

<File>   **<Save...>**

Command: **QSAVE**

2.6   Summary

In this chapter, you practiced the following AutoCAD commands, which are related to 3D solid modeling.

BOX	CONE	CYLINDER
SPHERE	TORUS	WEDGE
UNION	SUBTRACT	SLICE
CHAMFER	FILLET	MIRROR3D
ALIGN		

For a brief explanation of these commands, refer to the appendix of this book.

You have learned how to create primitives and how to create a complex solid by adding and subtracting simpler solids. In the next chapter, you will practice region modeling, the creation of solids by extruding and revolving a region or polyline, and the creation of complex solid models by intersecting solids.

2.7   Exercises

To further enhance your knowledge of using primitive solids to compose a complex solid model, complete the following exercises.

Figure 2.30 shows the engineering drawing views of an angle bracket. Before looking at Figure 2.31, which is the suggested breakdown, take some time to analyze the component to determine what primitive solids you need to make this object.

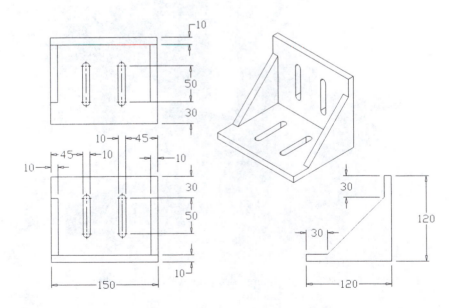

Figure 2.30 Engineering drawing of an angle bracket

Figure 2.31 is the suggested breakdown of the angle bracket. Refer to the dimensions shown in Figure 2.30 to create the primitives one by one and place them in a correct position relative to each other.

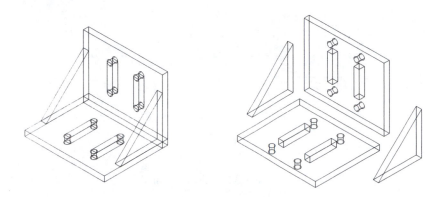

Figure 2.31 Suggested breakdown of the solid model

To form the complex solid, unite the two larger boxes and wedges. Then, subtract the remaining primitives. See Figure 2.32. Finally, fillet the edges A, B, C, D, E, F, G, H, J, K, and L.

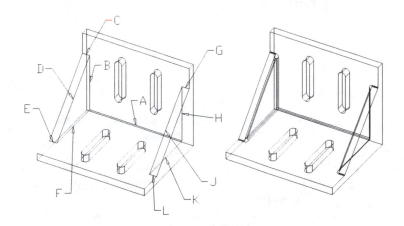

Figure 2.32 Eleven edges filleted

The angle bracket is complete. Save the drawing.

Start another new drawing to create the solid model of an adjusting pin, which is shown in Figure 2.33. Take some time to analyze the model to break it down into simple primitive solids. Then, compare your breakdown with Figure 2.34.

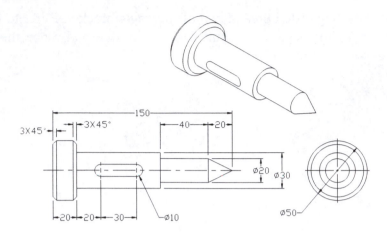

Figure 2.33 Adjusting pin

Figure 2.34 is the suggested breakdown of the pin. There are five solid cylinders, one solid cone, and one solid box.

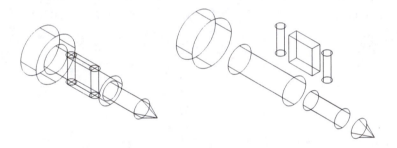

Figure 2.34 Suggested breakdown of the adjusting pin

Refer to the dimensions shown in Figure 2.33 to create the primitives. Then, unite the three larger cylinders and the solid cone. Next, subtract the remaining solids. Finally, chamfer two edges. See Figure 2.35.

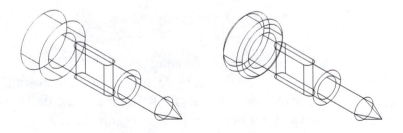

Figure 2.35 Two edges chamfered

The solid model is complete. Save your drawing.

Fillets and chamfers are commonly added to the edges of a solid model. When you need to fillet or chamfer a number of edges, you have to plan the sequence of work because the order in which the edges are treated affects the outcome. To appreciate this sequential effect, create three identical simple solid models and then round off the edges.

Start a new drawing and create three solid boxes. The corners of the first box are (0,0,0) and (50,50,50). The corners of the second box are (50,10,0) and (100,50,10). The corners of the third box are (50,40,0) and (100,50,20). Unite the three boxes. Then, array them for a distance of 150 units in the X direction. See Figure 2.36.

You will fillet the edges of these three models in different orders to see the effect of filleting sequence on the filleting outcome.

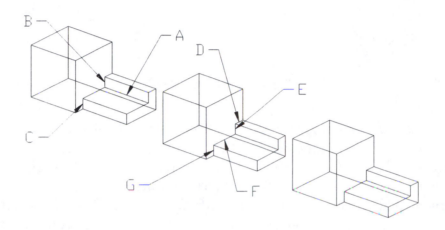

Figure 2.36 Three solid models

Use the FILLET command to fillet the edges A, B, and C of the left model, and to fillet the edges D, E, F, and G of the middle model. The fillet radius is 4 units. Compare your models with Figure 2.37.

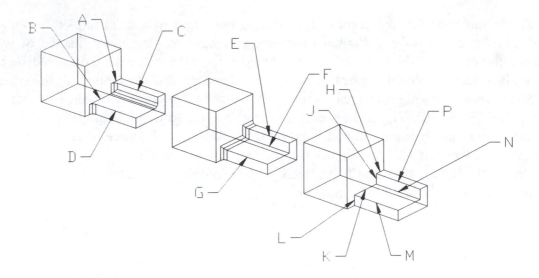

Figure 2.37 Edges filleted

To continue, fillet the edges H, J, K, L, M, N, and P of the right model in one operation. Then, in one operation again, fillet the edges A, B, C, and D of the left model, and the edges E, F, and G of the middle model. When you fillet the right and middle models, you have to select the CHAIN option to select the edges.

After running the command, you will find that the edges of left model cannot be filleted in one operation. If you leave out the edges C and D, the outcome should resemble Figure 2.38.

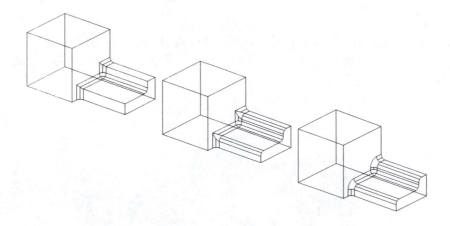

Figure 2.38 More edges filleted

As you can see in Figure 2.38, the outcome of the middle and the right models are different. Now, you can fillet the edges C and D again by using the CHAIN option. The left model should now be the same as the middle model.

From these results, you should realize that if you need to fillet a number of edges, you should be cautious about the sequence of fillet application.

Chapter 3
Extrude, Revolve, and Intersect

The basic building blocks available for constructing a complex solid model are boxes, cylinders, spheres, tori, and wedges. In the last chapter, you created complex solid models from them by using the Boolean operations, union and subtraction.

With these simple primitive solids, the types of complex solid that you can build are quite limited. To produce a wider repertoire of solid models, you can use extrusion and revolving. Both types of solids require a closed, 2D planar area that can be either a closed polyline or a region.

An extruded solid is formed by extruding a closed, 2D planar area in a direction perpendicular to the 2D plane. A revolved solid is formed by revolving a closed, 2D planar area about an axis.

In this chapter, you will carry out a number of exercises. First, you will practice region modeling, the creation of extruded solid and revolved solid from regions, and making the solid model of a hydraulic valve by intersection and subtraction. Next, you will produce the solid model of a watch case. This solid model is the intersection of an extruded solid and a revolved solid. Finally, you will retrieve the unsliced solid model created in the previous chapter, and learn how to apply the intersection process to cut it into two pieces along an irregular cutting edge.

3.1   A Different Approach

Figure 3.1 shows the completed solid model of a hydraulic valve. As depicted in the last chapter, you need to analyze the object in order to find out what primitives are required.

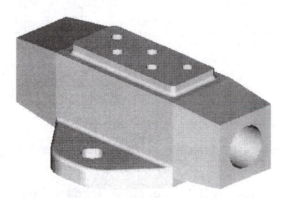

Figure 3.1 Hydraulic valve

This time, you will use a different approach. Instead of using primitives, you will make three extruded solids and perform intersection on them to yield the main body of the model. Then, you will create a solid of revolution and subtract it from the main body. Finally, you will fillet the edges and add five holes on the top face.

To make extruded solids, you need to produce closed planar regions or polylines. Figure 3.2 shows three profiles. They represent the top, front, and side views of the hydraulic valve. Although the diagram shows the three profiles separately in a single 2D plane, they are, in fact, lying on three planes: the top view lies on the XY plane, the front view lies on the XZ plane, and the side view lies on the YZ plane. The (0,0) positions on the profiles indicate the origin of the UCS.

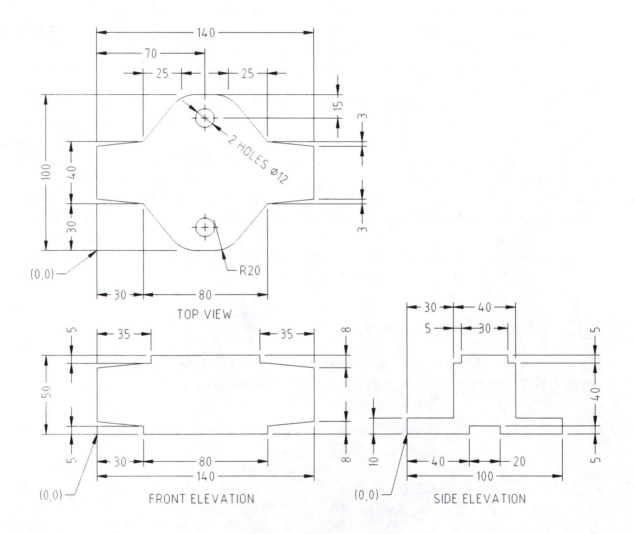

Figure 3.2 Profiles of the hydraulic valve main body

From these profiles, you will create three solids of extrusion. Figure 3.3 is an exploded view of the extruded solids.

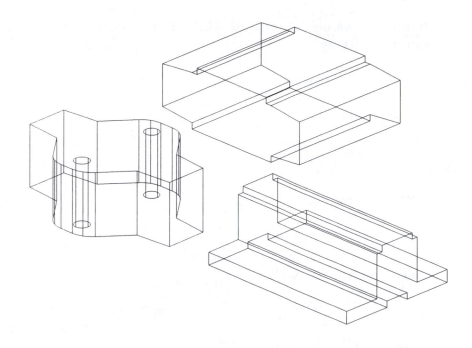

Figure 3.3 Extruded solids created from the profiles

The core of the hydraulic valve is made from a solid of revolution. Figure 3.4 shows the cross section and the revolved solid created from the cross section.

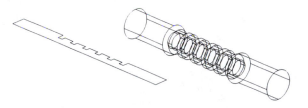

Figure 3.4 Cross section and the revolved solid

3.2   Region Modeling

A region is a closed, 2D planar area. You can produce a region from a combination of lines, polylines, circles, ellipse, elliptical arcs, and arcs. Just like with solid objects, you can perform Boolean operations on regions to create complex regions.

Use the NEW command to begin a new drawing. The prototype drawing is ACADISO.DWG.

    <File>   **<New...>**

    Command: **NEW**

Make three new layers called SPOOL, WIREFRAME and SOLID. The color of layer SPOOL is red, the color of layer WIREFRAME is cyan, and the color of layer SOLID is yellow. The current layer should be WIREFRAME. You will create the profiles for the three extruded solids on this layer.

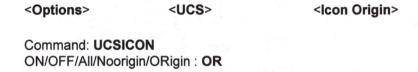

Command: **DDLMODES**

Layer	Color
SOLID	**yellow**
SPOOL	**red**
WIREFRAME	**cyan**

Current layer: **WIREFRAME**

Set the UCS icon on the origin position with the UCSICON command. This will give you a better idea of where you are working.

<**Options**>   <**UCS**>   <**Icon Origin**>

Command: **UCSICON**
ON/OFF/All/Noorigin/ORigin : **OR**

Figure 3.5 shows the profile for the top view, which you will make on the XY plane.

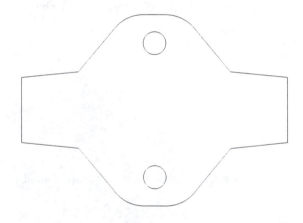

Figure 3.5 Profile of the top view

To create this profile, you will start by using the RECTANG command to create two rectangles. After that, use the ZOOM command to obtain a complete view of the rectangles. See Figure 3.6.

[**Polygon**]   [**Rectangle**]

Command: **RECTANG**
First corner: **0,30**
Other corner: **@140,40**

[**Polygon**]   [**Rectangle**]

Command: **RECTANG**
First corner: **30,0**
Other corner: **@80,100**

[**Zoom**]   [**Zoom Extents**]

Command: **ZOOM**
All/Center/Dynamic/Extents/Left/Previous/Vmax/Window/<Scale(X/XP)>: **E**
Regenerating drawing.

[**Zoom**]   [**Zoom Scale**]

Command: **ZOOM**
All/Center/Dynamic/Extents/Left/Previous/Vmax/Window/<Scale(X/XP)>: **.8X**

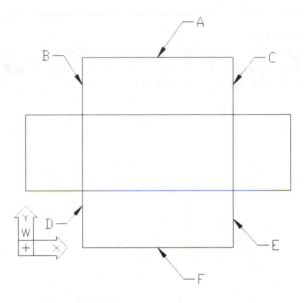

Figure 3.6 Two rectangles created

To obtain the required profile, you need to chamfer and fillet the rectangles. Use the CHAMFER command to set the chamfer distances to 25 and 30 units respectively, and to chamfer the corners of a rectangle. See Figure 3.7.

[**Feature**]   [**Chamfer**]

Command: **CHAMFER**
(TRIM mode) Current chamfer Dist1 = 0.0000, Dist2 = 0.0000
Polyline/Distance/Angle/Trim/Method/<Select first line>: D
Enter first chamfer distance: **25**
Enter second chamfer distance: **30**

[**Feature**]   [**Chamfer**]

Command: **CHAMFER**
(TRIM mode) Current chamfer Dist1 = 25.0000, Dist2 = 30.0000
Polyline/Distance/Angle/Trim/Method/<Select first line>: **[Select A (Figure 3.6).]**
Select second line: **[Select B (Figure 3.6).]**

[Feature]   **[Chamfer]**

Command: **CHAMFER**
(TRIM mode) Current chamfer Dist1 = 25.0000, Dist2 = 30.0000
Polyline/Distance/Angle/Trim/Method/<Select first line>: **[Select A (Figure 3.6).]**
Select second line: **[Select C (Figure 3.6).]**

[Feature]   **[Chamfer]**

Command: **CHAMFER**
(TRIM mode) Current chamfer Dist1 = 25.0000, Dist2 = 30.0000
Polyline/Distance/Angle/Trim/Method/<Select first line>: **[Select F (Figure 3.6).]**
Select second line: **[Select D (Figure 3.6).]**

[Feature]   **[Chamfer]**

Command: **CHAMFER**
(TRIM mode) Current chamfer Dist1 = 25.0000, Dist2 = 30.0000
Polyline/Distance/Angle/Trim/Method/<Select first line>: **[Select F (Figure 3.6).]**
Select second line: **[Select E (Figure 3.6).]**

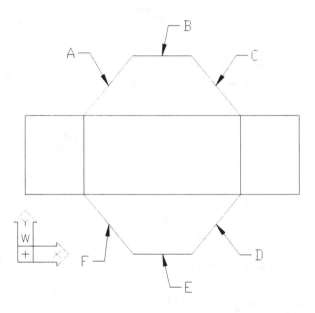

Figure 3.7 Rectangle chamfered

After chamfering, run the FILLET command to set the fillet radius to 20 units and to round off four corners. See Figure 3.8.

[Feature]   **[Fillet]**

Command: **FILLET**
(TRIM mode) Current fillet radius = 0.0000
Polyline/Radius/Trim/<Select first object>: **R**
Enter fillet radius: **20**

[Feature]   **[Fillet]**

Command: **FILLET**
(TRIM mode) Current fillet radius = 20.0000
Polyline/Radius/Trim/<Select first object>: **[Select A (Figure 3.7).]**
Select second object: **[Select B (Figure 3.7).]**

[Feature]   **[Fillet]**

Command: **FILLET**
(TRIM mode) Current fillet radius = 20.0000
Polyline/Radius/Trim/<Select first object>: **[Select B (Figure 3.7).]**
Select second object: **[Select C (Figure 3.7).]**

[Feature]   **[Fillet]**

Command: **FILLET**
(TRIM mode) Current fillet radius = 20.0000
Polyline/Radius/Trim/<Select first object>: **[Select D (Figure 3.7).]**
Select second object: **[Select E (Figure 3.7).]**

[Feature]   **[Fillet]**

Command: **FILLET**
(TRIM mode) Current fillet radius = 20.0000
Polyline/Radius/Trim/<Select first object>: **[Select E (Figure 3.7).]**
Select second object: **[Select F (Figure 3.7).]**

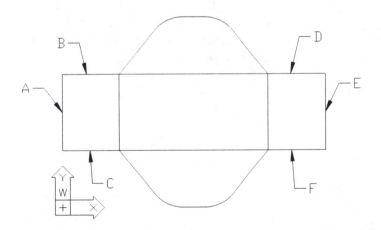

Figure 3.8 Four corners filleted

One of the rectangles is complete. To continue, run the CHAMFER command to set the chamfer distances to 3 and 30 respectively, and to chamfer the four corners of the other rectangle. See Figure 3.9.

[Feature]   **[Chamfer]**

Command: **CHAMFER**

(TRIM mode) Current chamfer Dist1 = 25.0000, Dist2 = 30.0000
Polyline/Distance/Angle/Trim/Method/<Select first line>: **D**
Enter first chamfer distance: **3**
Enter second chamfer distance: **30**

[Feature]   **[Chamfer]**

Command: **CHAMFER**
(TRIM mode) Current chamfer Dist1 = 3.0000, Dist2 = 30.0000
Polyline/Distance/Angle/Trim/Method/<Select first line>: **[Select A (Figure 3.8).]**
Select second line: **[Select B (Figure 3.8).]**

[Feature]   **[Chamfer]**

Command: **CHAMFER**
(TRIM mode) Current chamfer Dist1 = 3.0000, Dist2 = 30.0000
Polyline/Distance/Angle/Trim/Method/<Select first line>: **[Select A (Figure 3.8).]**
Select second line: **[Select C (Figure 3.8).]**

[Feature]   **[Chamfer]**

Command: **CHAMFER**
(TRIM mode) Current chamfer Dist1 = 3.0000, Dist2 = 30.0000
Polyline/Distance/Angle/Trim/Method/<Select first line>: **[Select E (Figure 3.8).]**
Select second line: **[Select D (Figure 3.8).]**

[Feature]   **[Chamfer]**

Command: **CHAMFER**
(TRIM mode) Current chamfer Dist1 = 3.0000, Dist2 = 30.0000
Polyline/Distance/Angle/Trim/Method/<Select first line>: **[Select E (Figure 3.8).]**
Select second line: **[Select F (Figure 3.8).]**

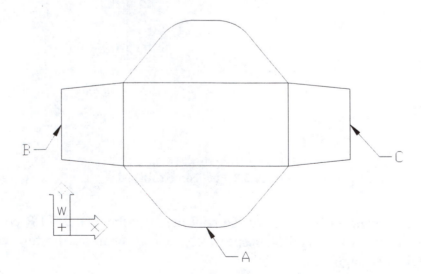

Figure 3.9 Second rectangle chamfered

The other rectangle is also complete. To proceed, use the CIRCLE command to create a circle of 6 units radius and use the MIRROR command to add another circle. See Figure 3.10.

[Draw]   **[Circle Center Radius]**

Command: **CIRCLE**
3P/2P/TTR/<Center point>: **FROM**
Base point: **MID** of **[Select A (Figure 3.9).]**
<Offset>: **@15<90**
Diameter/<Radius>: **6**

[Copy]   **[Mirror]**

Command: **MIRROR**
Select objects: **LAST**
Select objects: **[Enter]**
First point of mirror line: **MID** of **[Select B (Figure 3.9).]**
Second point: **MID** of **[Select C (Figure 3.9).]**
Delete old objects? <N> **N**

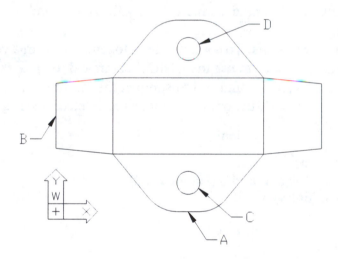

Figure 3.10 Two circles added

Now, you have four closed areas: two circles, a chamfered rectangle, and a chamfered and filleted rectangle. To form regions from them, you can run the REGION command. To reiterate, you can use any combination of lines, arcs, and polylines to define a closed area for subsequent conversion to a region. The main criteria for each single region is that the entities must form a single closed loop and must not cross each other. After converting the closed areas into regions, there should be no noticeable change on the screen. If you want to break a region down into lines and arcs, you can run the EXPLODE command.

Before you continue making the regions, you have to decide whether to retain or to delete the original object after it has been used. To delete the original object, you have to set the system variable DELOBJ to 1. If DELOBJ is 0, the original objects remain after an operation. Because the entities are not required after being used to make the regions, set the DELOBJ variable to 1.

Command: **DELOBJ**
New value for DELOBJ: **1**

With DELOBJ set to 1, the lines and arcs entities that were used to form a region will be deleted.
Run the REGION command.

[Polygon]   **[Region]**

Command: **REGION**
Select objects: **[Select A, B, C, and D (Figure 3.10).]**
Select objects: **[Enter]**
4 loops extracted.
4 Regions created.

As mentioned earlier, you will not find any noticeable change on the screen display. To know what kinds of entities are on the screen, you can use the LIST command. If you have use the REGION command correctly, the LIST command will tell you that the entities are regions.

Now, you have four regions. To combine them together to form a complex region that resembles Figure 3.5, you have to use the UNION command to unite the regions A and B, and then use the SUBTRACT command to subtract the regions C and D from the united region. Because DELOBJ is 1, the original regions are deleted. See Figure 3.11.

[Modify]   **[Union]**

Command: **UNION**
Select objects: **[Select A and B (Figure 3.10).]**
Select objects: **[Enter]**

[Modify]   **[Subtract]**

Command: **SUBTRACT**
Select solids and regions to subtract from...
Select objects: **[Select A (Figure 3.10).]**
Select objects: Select solids and regions to subtract...
Select objects: **[Select C and D (Figure 3.10).]**
Select objects: **[Enter]**

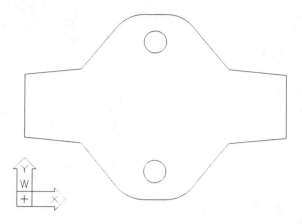

Figure 3.11 Boolean operations applied to the regions

The profile for the top view is completed. The second profile that you will be working on resides on the ZX plane of the WCS. To get a better view, set the display to a new viewing direction by using the VPOINT command. To set the new entities on the ZX plane, set the UCS to rotate 90° about the X axis by applying the UCS command. See Figure 3.12.

<View>   **<3D Viewpoint>**   **<Vector>**

Command: **VPOINT**
Rotate/<View point> <0.0000,0.0000,1.0000>: **R**
Enter angle in XY plane from X axis: **315**
Enter angle from XY plane: **25**
Regenerating drawing.

[UCS]   **[X Axis Rotate UCS]**

Command: **UCS**
Origin/ZAxis/3point/OBject/View/X/Y/Z/Prev/Restore/Save/Del/?/<World>: **X**
Rotation angle about X axis <0>: **90**

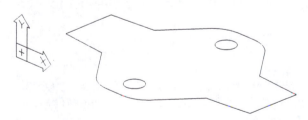

Figure 3.12 New viewing position and new UCS

Check your UCS icon against Figure 3.12. Then, use the RECTANG command to create three rectangles. See Figure 3.13.

[Polygon]   **[Rectangle]**

Command: **RECTANG**
First corner: **0,5**
Other corner: **@140,40**

[Polygon]   **[Rectangle]**

Command: **RECTANG**
First corner: **30,0**
Other corner: **@80,5**

[Polygon]   **[Rectangle]**

Command: **RECTANG**
First corner: **35,45**
Other corner: **@70,5**

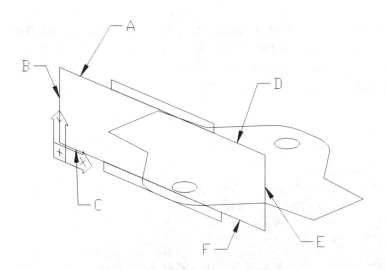

Figure 3.13 Three rectangles created on the new UCS

The large rectangle needs to be chamfered. Run the CHAMFER command to set the chamfer distances, and then repeat it four times to chamfer four corners. See Figure 3.14.

[Feature]   **[Chamfer]**

Command: **CHAMFER**
(TRIM mode) Current chamfer Dist1 = 3.0000, Dist2 = 30.0000
Polyline/Distance/Angle/Trim/Method/<Select first line>: **D**
Enter first chamfer distance: **3**
Enter second chamfer distance: **30**

[Feature]   **[Chamfer]**

Command: **CHAMFER**
(TRIM mode) Current chamfer Dist1 = 3.0000, Dist2 = 30.0000
Polyline/Distance/Angle/Trim/Method/<Select first line>: **[Select B (Figure 3.13).]**
Select second line: **[Select A (Figure 3.13).]**

[Feature]   [Chamfer]

Command: **CHAMFER**
(TRIM mode) Current chamfer Dist1 = 3.0000, Dist2 = 30.0000
Polyline/Distance/Angle/Trim/Method/<Select first line>: **[Select B (Figure 3.13).]**
Select second line: **[Select C (Figure 3.13).]**

[Feature]   [Chamfer]

Command: **CHAMFER**
(TRIM mode) Current chamfer Dist1 = 3.0000, Dist2 = 30.0000
Polyline/Distance/Angle/Trim/Method/<Select first line>: **[Select E (Figure 3.13).]**
Select second line: **[Select D (Figure 3.13).]**
[Feature]   [Chamfer]

Command: **CHAMFER**
(TRIM mode) Current chamfer Dist1 = 3.0000, Dist2 = 30.0000
Polyline/Distance/Angle/Trim/Method/<Select first line>: **[Select E (Figure 3.13).]**
Select second line: **[Select F (Figure 3.13).]**

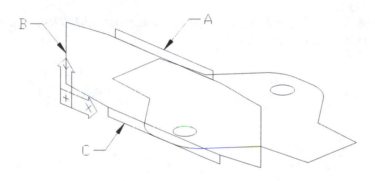

Figure 3.14 Larger rectangle chamfered

After chamfering, make three regions from the three closed areas by using the REGION command. Then, run the UNION command on the regions to combine them into a single region. See Figure 3.15.

[Polygon]   [Region]

Command: **REGION**
Select objects: **[Select A, B, and C (Figure 3.14).]**
Select objects: **[Enter]**
3 loops extracted.
3 Regions created.

[Modify]   [Union]

Command: **UNION**
Select objects: **[Select A, B, and C (Figure 3.14).]**
Select objects: **[Enter]**

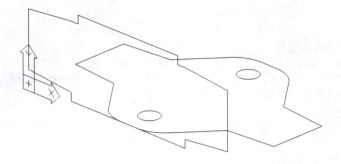

Figure 3.15 Regions united

The profile for the front elevation is complete. Before making the third region, rotate the UCS about the Y axis for 90°. The third profile resides on a new UCS plane. See Figure 3.16.

[UCS]   **[Y Axis Rotate UCS]**

Command: **UCS**
Origin/ZAxis/3point/OBject/View/X/Y/Z/Prev/Restore/Save/Del/?/<World>: **Y**
Rotation angle about Y axis <0>: **90**

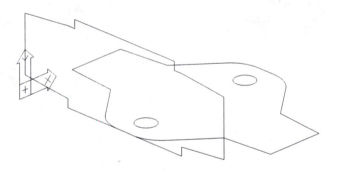

Figure 3.16 New UCS position

After setting the UCS, create three rectangles by using the RECTANG command. See Figure 3.17.

[Polygon]   **[Rectangle]**

Command: **RECTANG**
First corner: **0,0**
Other corner: **@40,10**

[Polygon]   **[Rectangle]**

Command: **RECTANG**
First corner: **30,5**
Other corner: **@40,40**

[Polygon]   **[Rectangle]**

Command: **RECTANG**
First corner: **35,45**
Other corner: **@30,5**

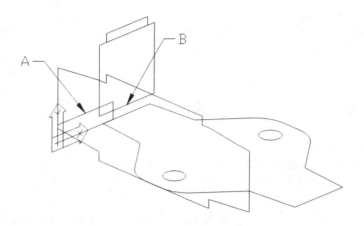

Figure 3.17 Three rectangles created

Use the MIRROR3D command to mirror a rectangle. See Figure 3.18.

[Copy]   **[3D Mirror]**

Command: **MIRROR3D**
Select objects: **[Select A (Figure 3.17).]**
Select objects: **[Enter]**
Plane by Object/Last/Zaxis/View/XY/YZ/ZX/<3points>: **YZ**
Point on YZ plane <0,0,0>: **MID** of **[Select B (Figure 3.17).]**
Delete old objects? <N> **N**

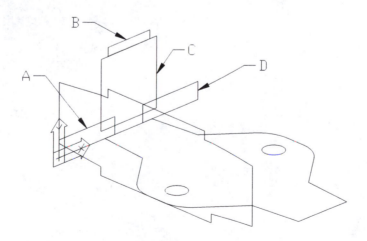

Figure 3.18 Rectangle mirrored

After mirroring the rectangle, convert the four rectangles to four regions by using the REGION command. Then, run the UNION command to unite the regions into a single region. See Figure 3.19.

[Polygon]   **[Region]**

Command: **REGION**
Select objects: **[Select A, B, C, and D (Figure 3.18).]**
Select objects: **[Enter]**
4 loops extracted.
4 Regions created.

[Modify]   **[Union]**

Command: **UNION**
Select objects: **[Select A, B, C, and D (Figure 3.18).]**
Select objects: **[Enter]**

The profiles for making three extruded solids are complete. To make a core, which is a revolved solid, you have to create the fourth region. Set a new UCS. See Figure 3.19.

[UCS]   **[World UCS]**

Command: **UCS**
Origin/ZAxis/3point/OBject/View/X/Y/Z/Prev/Restore/Save/Del/?/<World>: **W**

[UCS]   **[Origin UCS]**

Command: **UCS**
Origin/ZAxis/3point/OBject/View/X/Y/Z/Prev/Restore/Save/Del/?/<World>: **OR**
Origin point: **0,50,25**

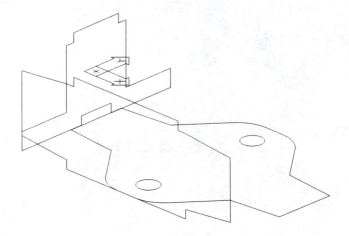

Figure 3.19 Four regions united and the UCS set to a new position

Set the current layer to SPOOL. You will put the wireframe for the spool on this layer.

<Data>   **<Layers...>**

Command: **DDLMODES**
Current layer: **SPOOL**

Run the PLAN command to set the viewing position to the plan view of the current UCS. Then, create a region using the dimensions shown in Figure 3.20.

<View>   **<3D Viewpoint Presets>**   **<Plan View>**   **<Current>**

Command: **PLAN**
<Current UCS>/Ucs/World: **C**
Regenerating drawing.

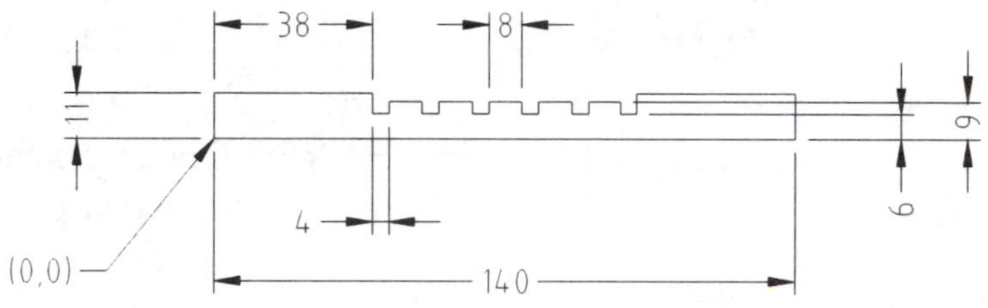

Figure 3.20 Dimensions for the fourth region

Use the RECTANG command and then the ARRAY command to create a series of rectangles. See Figure 3.21.

[Polygon]   **[Rectangle]**

Command: **RECTANG**
First corner: **0,0**
Other corner: **@140,11**

[Polygon]   **[Rectangle]**

Command: **RECTANG**
First corner: **38,9**
Other corner: **@64,10**

[Polygon]   **[Rectangle]**

Command: **RECTANG**
First corner: **38,6**
Other corner: **@4,15**

[Copy]   **[Rectangular Array]**

Command: **ARRAY**
Select objects: **LAST**
Select objects: **[Enter]**
Rectangular or Polar array (R/P) <R>: **R**
Number of rows (---) <1>: **1**
Number of columns (||||) <1>: **6**
Distance between columns (||||): **12**

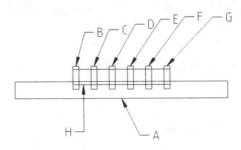

Figure 3.21 Eight rectangles created

Convert the rectangles into regions by using the REGION command. Then, use the SUBTRACT command to subtract the smaller rectangles from the large rectangle.

[Polygon]   **[Region]**

Command: **REGION**
Select objects: **[Select A, B, C, D, E, F, G, and H (Figure 3.21).]**
Select objects: **[Enter]**
8 loops extracted.
8 Regions created.

[Modify]   **[Subtract]**

Command: **SUBTRACT**
Select solids and regions to subtract from...
Select objects: **[Select A (Figure 3.21).]**
Select objects: Select solids and regions to subtract...
Select objects: **[Select B, C, D, E, F, G, and H (Figure 3.21).]**
Select objects: **[Enter]**

The four regions for making the complex solid are complete. Before starting to produce the solid model, set the viewing direction to 315° in the XY plane and 25° from the XY plane by using the VPOINT command. See Figure 3.22.

<View>   **<3D Viewpoint>**   **<Vector>**

Command: **VPOINT**
Rotate/<View point> <0.0000,0.0000,1.0000>: **R**
Enter angle in XY plane from X axis: **315**
Enter angle from XY plane: **25**
Regenerating drawing.

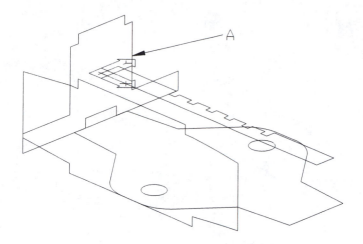

Figure 3.22 Fourth region created

3.3   Extrusion and Intersection

After completing the wireframes, you will start to build the solid model. To make the main body of the model, you will extrude three regions and then form an intersection of the extruded solids.

Set the current layer to SOLID. You will put the solid objects on this layer.

<Data>   **<Layers...>**

Command: **DDLMODES**
Current layer: **SOLID**

Set the system variable DELOBJ to 0 so that the region or polyline that is used for extrusion or revolving is retained. Retaining the original wireframes enables you to make the solid again quickly if anything goes wrong during solid creation.

Command: **DELOBJ**
New value for DELOBJ: **0**

Using the region on the ZY plane, run the EXTRUDE command to produce a solid of extrusion. See Figure 3.23.

[Solids]   **[Extrude]**

Command: **EXTRUDE**
Select objects: **[Select A (Figure 3.22).]**
Select objects: **[Enter]**
Path/<Height of Extrusion>: **140**
Extrusion taper angle: **0**

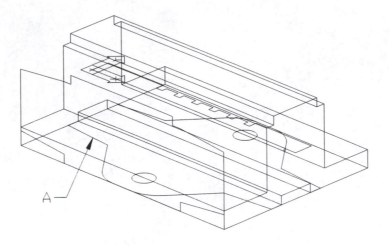

Figure 3.23 Region (side elevation) extruded

In specifying the height of extrusion, you need to aware that the direction, whether positive or negative, depends on the Z axis direction of the UCS on which the region or polyline is made, not the current UCS. Refer to Figure 3.18 to check the UCS orientation of the last extruded solid.

To make the second extruded solid, run the EXTRUDE command. See Figure 3.24.

[Solids]   **[Extrude]**

Command: **EXTRUDE**
Select objects: **[Select A (Figure 3.23).]**
Select objects: **[Enter]**
Path/<Height of Extrusion>: **50**
Extrusion taper angle: **0**

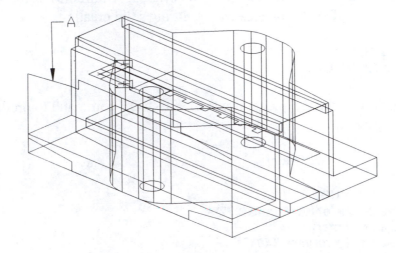

Figure 3.24 Region (top view) extruded

Apply the EXTRUDE command once more to create the third solid of extrusion. See Figure 3.25. To find out the UCS orientation of this region, refer to Figure 3.15. The direction of extrusion should be in the negative direction.

[Solids]   **[Extrude]**

Command: **EXTRUDE**
Select objects: **[Select A (Figure 3.24).]**
Select objects: **[Enter]**
Path/<Height of Extrusion>: **-100**
Extrusion taper angle: **0**

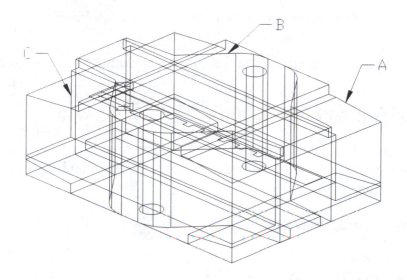

Figure 3.25 Region (front elevation) extruded

After making three extruded solids from the regions, run the INTERSECT command to produce a solid of intersection from them. A solid of intersection is a solid with the common volume of a set of solids. After intersection, turn off the layer WIREFRAME. See Figure 3.26.

[Modify]   **[Intersection]**

Command: **INTERSECT**
Select objects: **[Select A, B, and C (Figure 3.25).]**
Select objects: **[Enter]**

<Data>   <Layers...>

Command: **DDLMODES**

Layer	
WIREFRAME	**Off**

Current layer: **SOLID**

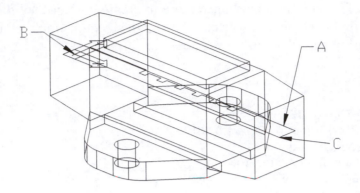

Figure 3.26 Intersection of three extruded solids created

The intersection of three extruded solids forms the main body of the hydraulic valve.

3.4   Revolving and Subtraction

To make the core of the hydraulic valve, you will revolve a region to form a solid of revolution and then subtract the revolved solid from the main body. Run the REVOLVE command. See Figure 3.27.

[Solids]   **[Revolve]**

Command: **REVOLVE**
Select objects: **[Select A (Figure 3.26).]**
Select objects: **[Enter]**
Axis of revolution - Object/X/Y/<Start point of axis>: **END** of **[Select B (Figure 3.26).]**
<End point of axis>: **END** of **[Select C (Figure 3.26).]**
Angle of revolution <full circle>: **[Enter]**

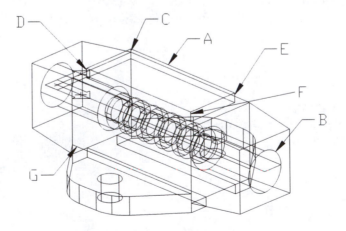

Figure 3.27 Fourth region revolved

After revolving the wireframe, it is not needed any more. Turn off the layer SPOOL. To cut a core on the main body using the revolved solid, run the SUBTRACT command.

Command: **DDLMODES**

Layer	
SPOOL	**Off**

Current layer: **SOLID**

[Modify]   **[Subtract]**

Command: **SUBTRACT**
Select solids and regions to subtract from...
Select objects: **[Select A (Figure 3.27).]**
Select objects: Select solids and regions to subtract...
Select objects: **[Select B (Figure 3.27).]**
Select objects: **[Enter]**

3.5   Filleting and Adding Holes

To complete the hydraulic valve, you have to round off some corners and cut five holes.

Create a series of fillets on the solid model by using the FILLET command. See Figure 3.28.

[Feature]   **[Fillet]**

Command: **FILLET**
(TRIM mode) Current fillet radius = 0.0000
Polyline/Radius/Trim/<Select first object>: **[Select C (Figure 3.27).]**
Enter radius: **2**
Chain/Radius/<Select edge>: **[Select D, E, F, and G (Figure 3.27).]**
Chain/Radius/<Select edge>: **[Enter]**
5 edges selected for fillet.

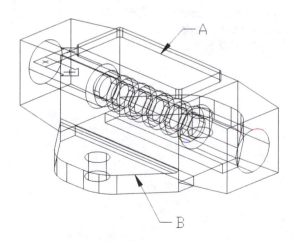

Figure 3.28 Five edges filleted

Run the FILLET command to make two chains of fillets. See Figure 3.29.

[**Feature**]   [**Fillet**]

Command: **FILLET**
(TRIM mode) Current fillet radius = 2.0000
Polyline/Radius/Trim/<Select first object>: **[Select A (Figure 3.28).]**
Enter radius: **2**
Chain/Radius/<Select edge>: **C**
Edge/Radius/<Select edge chain>: **[Select A and B (Figure 3.28).]**
Edge/Radius/<Select edge chain>: **[Enter]**
15 edges selected for fillet.

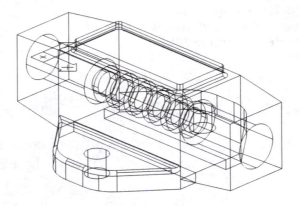

Figure 3.29 Two chains of edges filleted

Before filleting the other side of the valve, run the VPOINT command to set to a new viewing position. See Figure 3.30.

<View>   <3D Viewpoint>   <Vector>

Command: **VPOINT**
Rotate/<View point> <-0.6409,0.6409,0.4226>: **R**
Enter angle in XY plane from X axis: **135**
Enter angle from XY plane: **25**
Regenerating drawing.

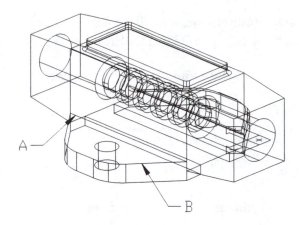

Figure 3.30 New viewing position

Run the FILLET command again. Then, set the UCS back to WORLD by using the UCS command. See Figure 3.31.

[Feature]   **[Fillet]**

Command: **FILLET**
(TRIM mode) Current fillet radius = 2.0000
Polyline/Radius/Trim/<Select first object>: **[Select A (Figure 3.30).]**
Enter radius: **2**
Chain/Radius/<Select edge>: **C**
Edge/Radius/<Select edge chain>: **[Select B (Figure 3.30).]**
Edge/Radius/<Select edge chain>: **[Enter]**
8 edges selected for fillet.

[UCS]   **[World UCS]**

Command: **UCS**
Origin/ZAxis/3point/OBject/View/X/Y/Z/Prev/Restore/Save/Del/?/<World>: **W**

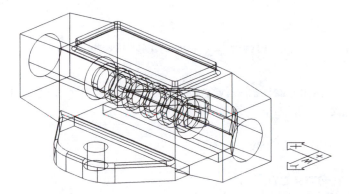

Figure 3.31 Edges of the other side filleted

Set the display to the previous viewing direction by using the VPOINT command. Then, create five cylinders by using the CYLINDER command. See Figure 3.32.

<View>   **<3D Viewpoint>**   **<Vector>**

Command: **VPOINT**
Rotate/<View point> <-0.6409,0.6409,0.4226>: **R**
Enter angle in XY plane from X axis: **315**
Enter angle from XY plane: **25**
Regenerating drawing.

[Solids]   **[Cylinder]**   **[Center]**

Command: **CYLINDER**

Center location	Radius	Height
46,56,25	**3**	**40**
70,56,25	**3**	**40**
94,56,25	**3**	**40**
58,44,25	**3**	**40**
82,44,25	**3**	**40**

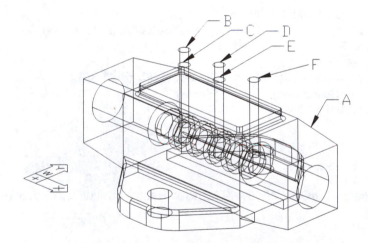

Figure 3.32 Five cylinders created

To cut the holes on the valve body, run the SUBTRACT command to subtract the five cylinders. See Figure 3.33.

[Modify]   **[Subtract]**

Command: **SUBTRACT**
Select solids and regions to subtract from...
Select objects: **[Select A (Figure 3.32).]**
Select objects: Select solids and regions to subtract...
Select objects: **[Select B, C, D, E, and F (Figure 3.32).]**
Select objects: **[Enter]**

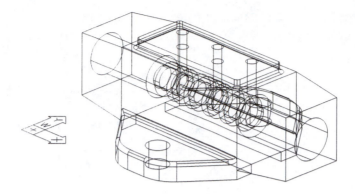

Figure 3.33 Completed hydraulic valve

The solid model of the hydraulic valve is complete.

3.6   Further Practice

For further practice on making an extruded solid, a revolved solid, and a solid of intersection, you will create the solid model of a watch case. See Figure 3.34.

Figure 3.34 Watch case

To make this solid model, you need to prepare two profiles that represent the top view and the cross section. See Figure 3.35. From the profiles, you will produce two regions. The region that is made from the top view will be used to create an extruded solid. The region that is made from the cross section will be used to create a revolved solid. The final complex solid model of the watch case is the intersection of these two solids.

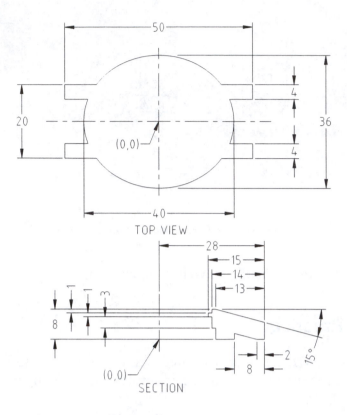

Figure 3.35 Profiles of the watch case

Use the NEW command to start a new drawing. Use ACADISO.DWG as the prototype.

<File>   **<New...>**

Command: **NEW**

Prepare your drawing. First, create a new layer called SOLID, and set it to current with the LAYER or DDLMODES command. Then, place the UCS icon at the origin position by using the UCSICON command. See Figure 3.36. Create the required entities by using the ELLIPSE, LINE, and TRIM commands. Next, convert them to a closed polyline using the PEDIT command, or to a region using the REGION command.

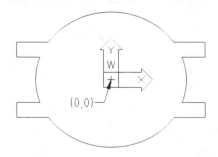

Figure 3.36 Profile of the top view created

Set the display to a new viewing position by using the VPOINT command. Then, set UCS to rotate 90° about the X axis. See Figure 3.37.

<View>   **<3D Viewpoint>**   **<Vector>**

Command: **VPOINT**
Rotate/<View point>: **R**
Enter angle in XY plane from X axis: **315**
Enter angle from XY plane: **25**
Regenerating drawing.

[UCS]   **[X Axis Rotate UCS]**

Command: **UCS**
Origin/ZAxis/3point/OBject/View/X/Y/Z/Prev/Restore/Save/Del/?/<World>: **X**
Rotation angle about X axis: **90**

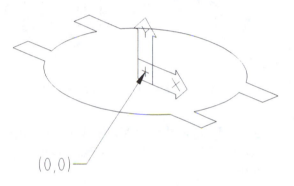

Figure 3.37 UCS set to a new position

Referring to Figure 3.35 and Figure 3.38, create another set of entities. Then, convert them to either a polyline or a region.

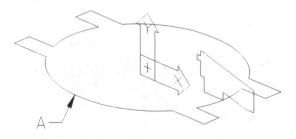

Figure 3.38 Profile of the cross section created

The wireframes for making the solid model of the watch case are completed. The solid model is an intersection of two solids — an extruded solid and a revolved solid. To produce the extruded solid, run the EXTRUDE command. See Figure 3.39.

[Solids]   **[Extrude]**

Command: **EXTRUDE**
Select objects: **[Select A (Figure 3.38).]**
Select objects: **[Enter]**
Path/<Height of Extrusion>: **8**
Extrusion taper angle: **0**

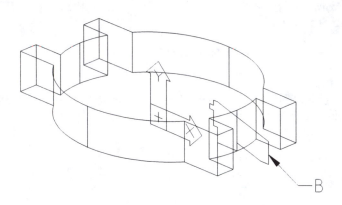

Figure 3.39 Extruded solid created

To produce the revolved solid, issue the REVOLVE command. See Figure 3.40.

[Solids]   **[Revolve]**

Command: **REVOLVE**
Select objects: **[Select B (Figure 3.39).]**
Select objects: **[Enter]**
Axis of revolution - Object/X/Y/<Start point of axis>: **Y**
Angle of revolution <full circle>: **[Enter]**

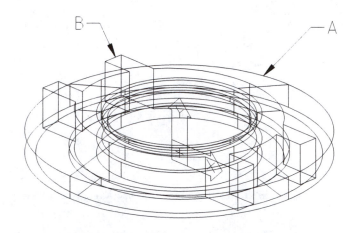

Figure 3.40 Revolved solid created

With the extruded solid and the revolved solid ready, run the INTERSECT command to form a solid of intersection. See Figure 3.41.

[Modify]   **[Intersection]**

Command: **INTERSECT**
Select objects: **[Select A and B (Figure 3.40).]**
Select objects: **[Enter]**

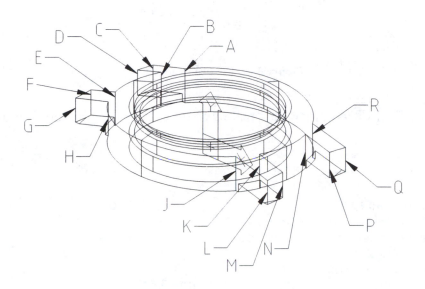

Figure 3.41 Intersection of the revolved solid and the extruded solid created

The main body of the watch case is completed. To round off the edges, run the FILLET command. See Figure 3.42.

[Feature]   **[Fillet]**

Command: **FILLET**
(TRIM mode) Current fillet radius = 0.0000
Polyline/Radius/Trim/<Select first object>: **[Select A (Figure 3.41).]**
Enter radius: **2**
Chain/Radius/<Select edge>: **[Select A, B, C, D, E, F, G, H, J, K, L, M, N, P, Q, and R (Figure 3.41).]**
Chain/Radius/<Select edge>: **[Enter]**
16 edges selected for fillet.

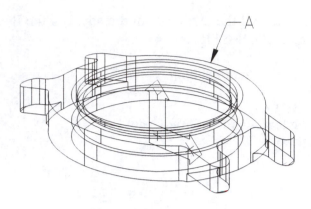

Figure 3.42 Sixteen edges filleted

Repeat the FILLET command to create a chain of fillets on the top face of the model. See Figure 3.43.

[Feature]   [Fillet]

Command: **FILLET**
(TRIM mode) Current fillet radius = 0.0000
Polyline/Radius/Trim/<Select first object>: **[Select A (Figure 3.42).]**
Enter radius: **2**
Chain/Radius/<Select edge>: **C**
Chain/Radius/<Select edge>: **[Select A (Figure 3.42).]**
Chain/Radius/<Select edge>: **[Enter]**
28 edges selected for fillet.

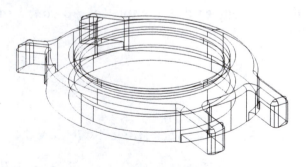

Figure 3.43 Complex solid model of the watch case

The solid model of the watch case is complete.

3.7   Solid Cutting by Intersection

In the final part of the previous chapter, you saved the model in two files. In one drawing file, you sliced the solid model into two pieces. Because the SLICE command gives a straight cutting edge, the cut edge is straight and flat.

Now, you will retrieve the saved file of the uncut solid model and cut it into two pieces with an irregular cutting edge by intersection.

<File>   **<Open...>**

Command: **OPEN**

When the dialog box appears, select the file of the uncut solid. See Figure 2.26.
Set the UCS to WORLD, and then rotate it for 90° about the X axis. See Figure 3.44.

[UCS]   **[World UCS]**

Command: **UCS**
Origin/ZAxis/3point/OBject/View/X/Y/Z/Prev/Restore/Save/Del/?/<World>: **W**

[UCS]   **[X Axis Rotate UCS]**

Command: **UCS**
Origin/ZAxis/3point/OBject/View/X/Y/Z/Prev/Restore/Save/Del/?/<World>: **X**
Rotation angle about X axis: **90**

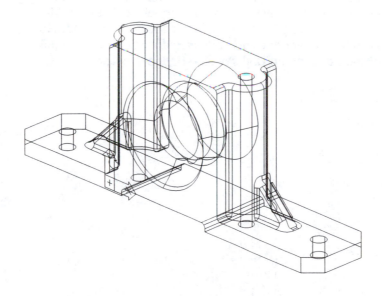

Figure 3.44 Uncut solid created in the previous chapter

In order to make a cut solid with an irregular cutting edge, you need to create an extruded solid with an irregular edge and intersect the extruded solid with the existing solid. After intersection, a portion of the original solid will be removed. To cut a solid and to keep both sides of the solid, you have to make a copy of the original solid, make two extruded solids, and intersect the two extruded solids with the two copies of the original solid.

To prepare two extruded solid, create two polylines by using the PLINE command. See Figure 3.45.

[Draw]   **[Polyline]**

Command: **PLINE**
From point: **-85,-10**
Current line-width is 0.0000
Arc/Close/Halfwidth/Length/Undo/Width/<Endpoint of line>: **-85,50**
Arc/Close/Halfwidth/Length/Undo/Width/<Endpoint of line>: **10,50**
Arc/Close/Halfwidth/Length/Undo/Width/<Endpoint of line>: **10,45**
Arc/Close/Halfwidth/Length/Undo/Width/<Endpoint of line>: **80,45**
Arc/Close/Halfwidth/Length/Undo/Width/<Endpoint of line>: **80,50**
Arc/Close/Halfwidth/Length/Undo/Width/<Endpoint of line>: **175,50**
Arc/Close/Halfwidth/Length/Undo/Width/<Endpoint of line>: **175,-10**
Arc/Close/Halfwidth/Length/Undo/Width/<Endpoint of line>: **C**

[Draw]   **[Polyline]**

Command: **PLINE**
From point: **-85,95**
Current line-width is 0.0000
Arc/Close/Halfwidth/Length/Undo/Width/<Endpoint of line>: **-85,50**
Arc/Close/Halfwidth/Length/Undo/Width/<Endpoint of line>: **10,50**
Arc/Close/Halfwidth/Length/Undo/Width/<Endpoint of line>: **10,45**
Arc/Close/Halfwidth/Length/Undo/Width/<Endpoint of line>: **80,45**
Arc/Close/Halfwidth/Length/Undo/Width/<Endpoint of line>: **80,50**
Arc/Close/Halfwidth/Length/Undo/Width/<Endpoint of line>: **175,50**
Arc/Close/Halfwidth/Length/Undo/Width/<Endpoint of line>: **175,95**
Arc/Close/Halfwidth/Length/Undo/Width/<Endpoint of line>: **C**

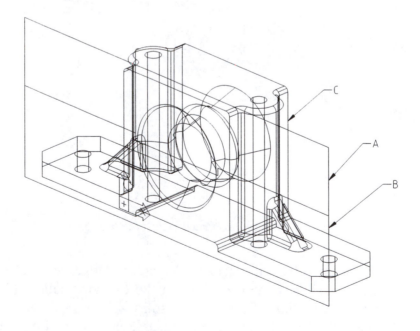

Figure 3.45 Two polylines created

Make a copy of the original solid by using the COPY command. Because the copied solid and the original solid are at the same position, there should no noticeable change on the screen.

[Modify]   **[Copy Object]**

Command: **COPY**
Select objects: **[Select C (Figure 3.45).]**
Select objects: **[Enter]**
<Base point or displacement>/Multiple: **0,0**
Second point of displacement: **[Enter]**

To produce two extruded solid from the two polylines, run the EXTRUDE command. See Figure 3.46.

[Solids]   **[Extrude]**

Command: **EXTRUDE**
Select objects: **[Select A and B (Figure 3.45).]**
Select objects: **[Enter]**
Path/<Height of Extrusion>: **-50**
Extrusion taper angle: **0**

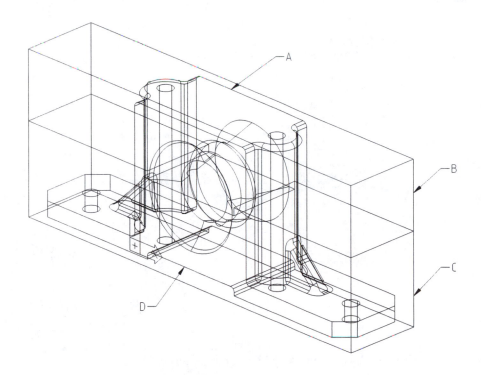

Figure 3.46 Two extruded solids created

Now, you have two copies of the original solid and two extruded solids. To cut one copy of the original solid, run the INTERSECT command. Select the upper extruded solid and the upper part of the complex solid. You will obtain the upper, cut half of the model.

[Modify]   **[Intersection]**

Command: **INTERSECT**
Select objects: **[Select A and B (Figure 3.46).]**
Select objects: **[Enter]**

After intersecting, the lower half of a copy of the original complex solid is removed. Because you have two copies of the original solid, you will use the second copy to intersect the lower extruded solid. The result is that you have two parts created, each one from a different copy of the solid.

Refresh the screen using the REDRAW command, and then apply the INTERSECT command once again. This time, select the lower extruded solid and the lower part of the complex solid. See Figure 3.47.

[Standard Toolbar]   **[Redraw View]**

Command: **REDRAW**

[Modify]   **[Intersection]**

Command: **INTERSECT**
Select objects: **[Select C, and D (Figure 3.46).]**
Select objects: **[Enter]**

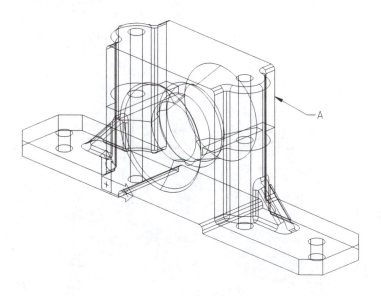

Figure 3.47 Complex solid cut into two pieces by intersection

In order to see clearly the effect of cutting, run the MOVE command to separate the upper piece. See Figure 3.48.

[Modify]   **[Move]**

Command: **MOVE**

Select objects: **[Select A (Figure 3.47).]**
Select objects: **[Enter]**
Base point or displacement: **45,45**
Second point of displacement: **[Enter]**

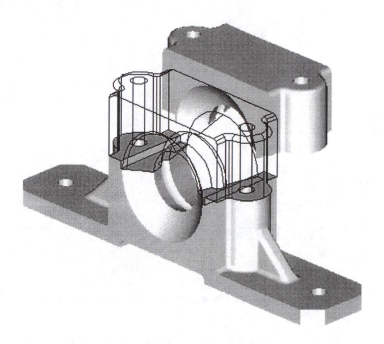

Figure 3.48 Cut model moved apart

3.8 Summary

In this chapter, you practiced the following AutoCAD commands related to 3D solid modeling.

REGION   EXTRUDE   REVOLVE
INTERSECTION

For a brief explanation of these commands, refer to the appendix of this book.

By now, you should have gained a thorough understanding of how to create regions and 2D profiles, how to produce extruded solids and revolved solids from 2D profiles, and how to use the intersection process in model creation. In the next chapter, you will learn how to create a set of thin shell solid models with internal bosses and webs, and run the utility commands on the solids.

3.9 Exercises

To enhance your knowledge of solid modeling, you will complete the following exercises on your own.

Figure 3.49 shows the engineering drawing of the solid model of the lower suspension arm of a scale model car. Take some time to analyze the solid model to determine what primitives and Boolean operations you would use.

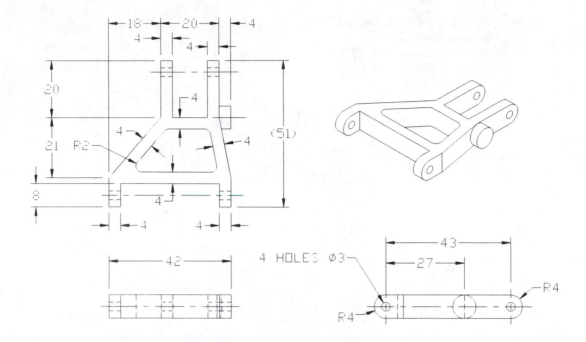

Figure 3.49 Lower suspension arm of a scale model car

Figure 3.50 is a suggested way to make the model. Create two regions and extrude them to form solids of extrusion. Then, make a number of solid cylinders. After making all the necessary primitive solids, use Boolean operations to compose the complex solid model. Save your drawing.

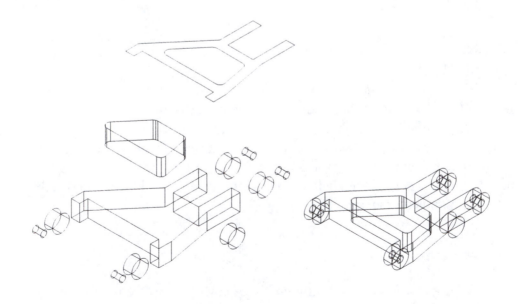

Figure 3.50 Suggested breakdown of the lower suspension arm

Start a new drawing. Create a solid model of the upper suspension arm of a model car. See Figure 3.51. This model, though similar to the lower arm that you created, can be produced by using a different approach. Instead of using the union operation, you will use the intersection operation. Before looking at Figure 3.52, the suggested breakdown, think for a moment how you can apply the intersection operation.

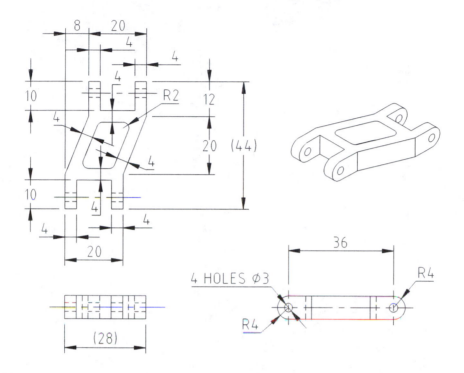

Figure 3.51 Upper suspension arm of the scale model car

A suggested breakdown of the model is shown in Figure 3.52. You can produce two sets of wireframes, and create the main body of the model by extrusion and then intersection. After intersection, add four holes. Save your work.

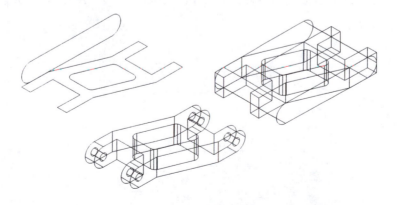

Figure 3.52 Suggested breakdown of the upper suspension arm

Start a new drawing to produce the solid model of the U-bracket of the model car. See Figure 3.53. Analyze the model to find out what primitive solids are required. Then, compare your analysis with the suggested breakdown shown in Figure 3.54.

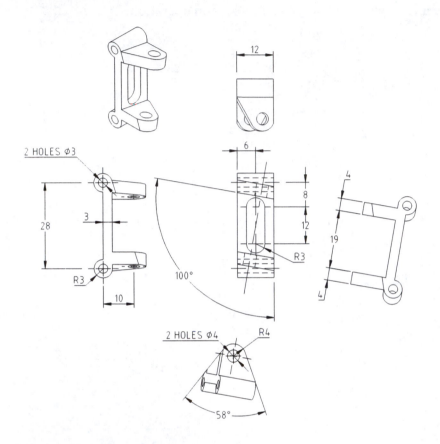

Figure 3.53 U-bracket of the scale model car

Figure 3.54 shows the suggested breakdown of the U-bracket. To make the model, you can start by making the primitives shown on the left of the figure. Then, you have to rotate the two triangular pieces and compose the solid model by using union and subtraction. To complete the model, you can make a large solid box and perform intersection.

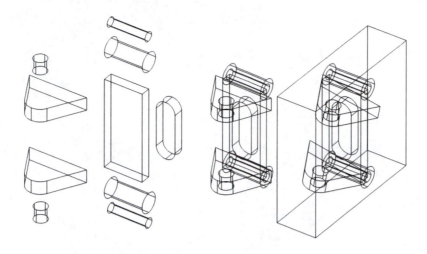

Figure 3.54 The suggested breakdown of the U-bracket

Start a new drawing to create a solid model for the transmission box of a model car. Read Figure 3.55 carefully, and prepare a breakdown of the model.

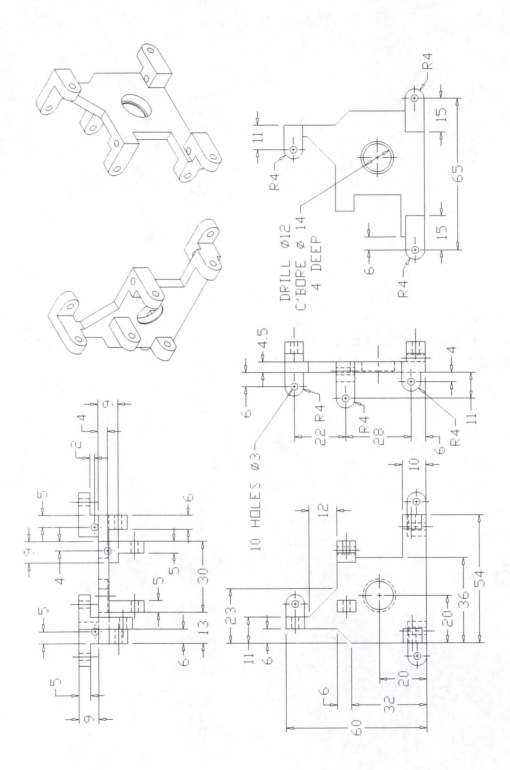

Figure 3.55 Half transmission box of the scale model car

Figure 3.56 shows the suggested breakdown of the model. The main body is a solid of extrusion. Other features are a solid box and solid cylinders that are either united to or subtracted from the main body. Create the primitives accordingly, and compose them together. Save your work.

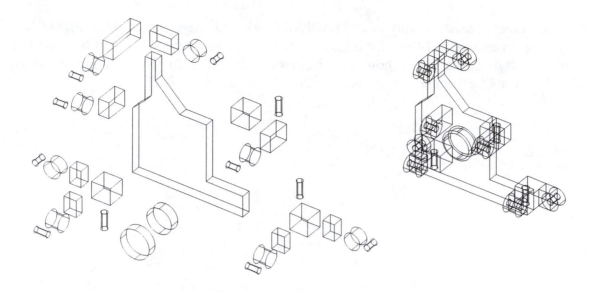

Figure 3.56 Suggested breakdown of the transmission box

Start a new drawing to produce the solid model of the rear hub of a model car. See Figure 3.57. Take a moment to analyze its constituents. Then, compare your breakdown with Figure 3.58.

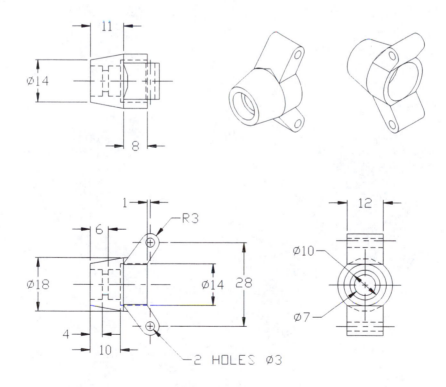

Figure 3.57 Rear hub of the scale model car

The model consists mainly of two revolved solids and one extruded solid. Figure 3.58 shows the dimensions of the wireframes for making the revolved solids and the extruded solid. The figure also shows how the wireframes relate to each other. Create the solids accordingly, and compose the complex model. Then, save your work.

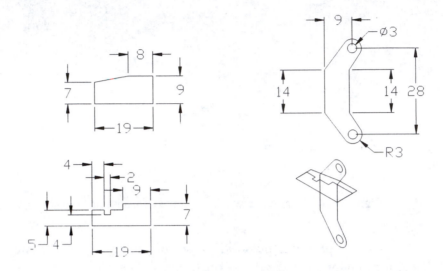

Figure 3.58 Wireframes for making the rear hub

Chapter 4
Product Design Project

After learning the basic skills in model creation using constructive solid geometry techniques, you will build a set of solid models for the casing of an electro-mechanical product in this chapter. Like many domestic appliances, this product consists of electronic and mechanical components that are enclosed within a polymeric casing. The casing consists of three parts: the upper casing, the lower casing, and the battery cover. See Figure 4.1. In order for you to have a better idea of how the three solid models look like, Figure 4.2 shows an exploded view of the product.

The main aim of working on this project is, in addition to enhancing your knowledge, to gain an appreciation of making a 3D thin shell solid model with internal webs and bosses.

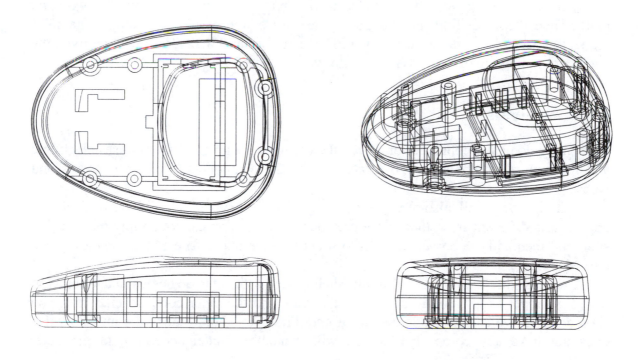

Figure 4.1 Casing of the electro-mechanical product

81

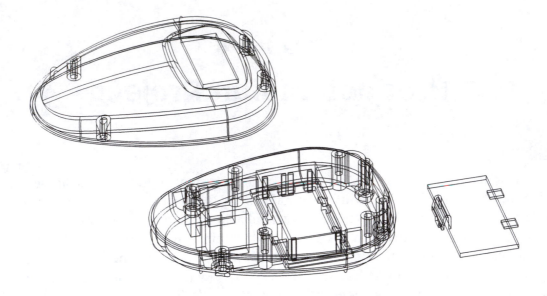

Figure 4.2 Upper casing, lower casing, and battery cover

In the following delineation, you will be guided to complete the upper and lower casings. After making the two solid models, you will put them together to check if there is any interference between them, and you will create sections to see how the two parts mate. Then, you will find out the mass properties of the models and output the solid models to Stereolithography format and 3D Studio format. Finally, you will be given the engineering drawing of the battery cover for you to complete the assembly.

4.1   Analysis

Like many domestic consumer products, the casing is a thin shell that houses electronic and mechanical components. To locate and fix the components, there are bosses and webs.

Although you will also use the building block principle to build the model, the approach is different from that of the previous projects. You cannot simply make a thin shell and then add the bosses or webs to it because the upper face of the top casing is a curved surface,

A general direction to model a thin shell object with internal bosses and webs is to treat the model as two solid parts: the outer skin, and the inner core. To elaborate, you should make two solid models instead of one. First, you should make the model as if it does not have any core. That is, you will make the model according to the outer dimensions. See Figure 4.3.

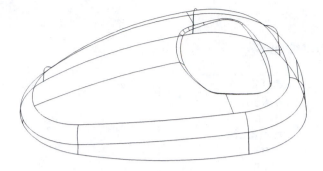

Figure 4.3 Solid that represents the outer skin of the upper casing

Then, you should model the inner core. To make the inner core, you should regard the inner core as a solid object, not a void. In this sense, the webs and bosses on the final model should appear as recesses in the core solid. See Figure 4.4.

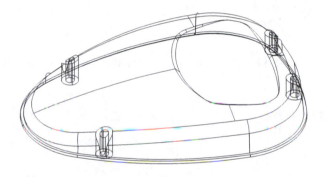

Figure 4.4 Solid that represents the core of the upper casing

After you have made both the outer skin and the inner core as solid objects, you can then subtract the inner core solid from the outer skin solid to yield the final model.

Basically, the method to create the lower casing is similar to that of the upper casing. You will carry out model creation in three steps. First, you will make the outer skin, then the inner core. Finally, you will subtract the inner core from the outer skin.

4.2   Drawing Preparation

To work systematically, you will use the DDLMODES command to create six layers — W_U, W_L, W_B, S_U, S_L1 and S_L2.

You will use layer W_U for the wireframes of the upper casing, layer W_L for the wireframes of the lower casing, layer W_B for the wireframes of the battery box, layer S_U for the solid model of the upper casing, and layers S_L1 and S_L2 for the solid model of the lower casing.

Also set the current layer to W_U. You will start by working on the wireframes for the upper casing.

<Data>   <Layers...>

Command: **DDLMODES**

Layer	Color
S_L1	green
S_L2	blue
S_U	magenta
W_B	red
W_L	cyan
W_U	yellow

Current layer: **W_U**

As a routine for 3D model creation, set the UCS icon to display at the origin position by using the UCSICON command.

<Options>   <UCS>   <Icon Origin>

Command: **UCSICON**
ON/OFF/All/Noorigin/ORigin: **OR**

Set the system variable DELOBJ to 1 so that the entities that are used to make the regions are deleted.

Command: **DELOBJ**
New value for DELOBJ: **1**

4.3   Wireframes for the Upper Casing

You can divide the wireframes for the upper casing into four groups. Figures 4.5 through Figure 4.8 delineates their X and Y coordinates.

Create three sets of wireframes. Then, convert them to form regions, as shown in Figure 4.5.

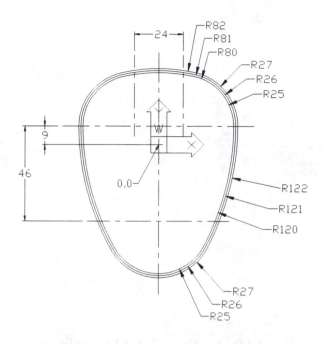

Figure 4.5 Three regions created

Create two more regions as shown in Figure 4.6. After that, move the outer region a distance of 13 units in the Z direction, and move the inner region a distance of 15 units in the Z direction.

[Modify]   **[Move]**

Command: **MOVE**
Select objects: **[Select A (Figure 4.6).]**
Select objects: **[Enter]**
Base point or displacement: **0,0,13**
Second point of displacement: **[Enter]**

[Modify]   **[Move]**

Command: **MOVE**
Select objects: **[Select B (Figure 4.6).]**
Select objects: **[Enter]**
Base point or displacement: **0,0,15**
Second point of displacement: **[Enter]**

To better see the position of the entities, set the display to two-viewport, and set the right viewport to an isometric view by setting the viewing direction to 315° in the XY plane and 25° from the XY plane.

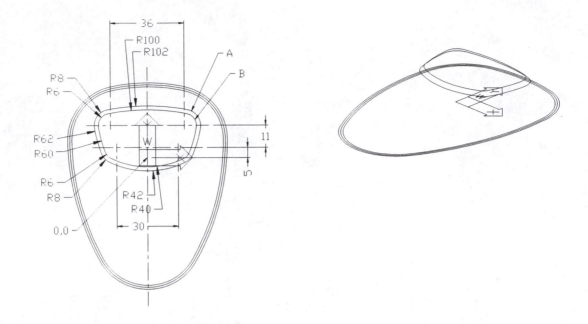

Figure 4.6 Two more regions created and moved

Run the CIRCLE command to create eight circles. Then, move four circles A, B, C, and D a distance of 6 units in the Z direction. See Figure 4.7.

[Modify]   **[Move]**

Command: **MOVE**
Select objects: **[Select A, B, C, and D (Figure 4.7).]**
Select objects: **[Enter]**
Base point or displacement: **0,0,6**
Second point of displacement: **[Enter]**

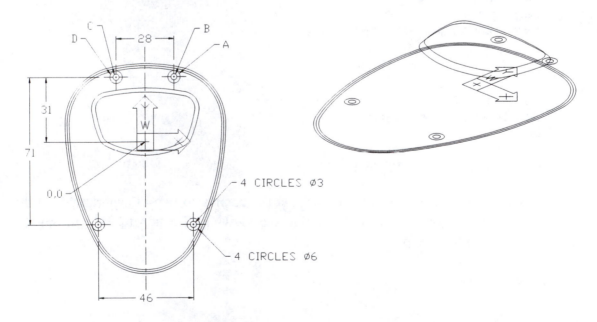

Figure 4.7 Eight circles created and four circles moved

Create a region and move it a distance of 11 units in the Z direction. See Figure 4.8.

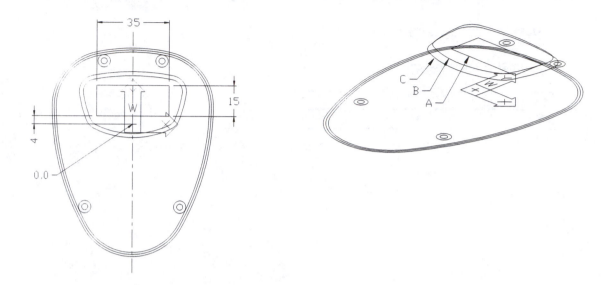

Figure 4.8 Rectangular region created and moved

The wireframe for the upper casing is complete.

4.4   Wireframes of the Lower Casing

The lower casing and the upper casing are two complementary parts that assemble together. Therefore, some of the profiles, and of course the wireframes, will be the same for both the upper casing and the lower casing.

In order to reduce the effort needed to make the wireframes, run the COPY command to copy some of the wireframes from the upper casing. After copying, change them to layer W_L.

[Modify]   **[Copy Object]**

Command: **COPY**
Select objects: **[Select all the entities except A, B, and C (Figure 4.8).]**
Select objects: **[Enter]**
<Base point or displacement>/Multiple: **0,0**
Second point of displacement: **[Enter]**

<Edit>   **<Properties...>**

Select objects: **P**
Select objects: **[Enter]**

[Properties
Layer...   **W_L**
OK   ]

For the time being, the wireframes for the upper casing is not required. Set the current layer to W_L and turn off the layer W_U. See Figure 4.9.

            <Data>   **<Layers...>**

Command: **DDLMODES**

Layer	
W_U	**Off**

Current layer: **W_L**

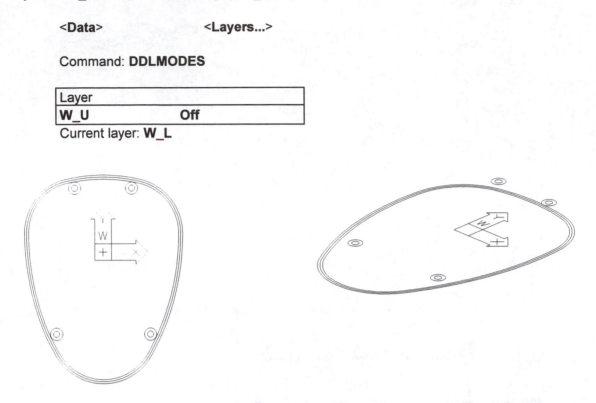

Figure 4.9 Eleven entities copied

As shown in Figure 4.10, add two circles. Then, move them a distance of -12.5 units in the Z direction.

        [Modify]   **[Move]**

Command: **MOVE**
Select objects: **[Select the two newly created circles (Figure 4.10).]**
Select objects: **[Enter]**
Base point or displacement: **0,0,-12.5**
Second point of displacement: **[Enter]**

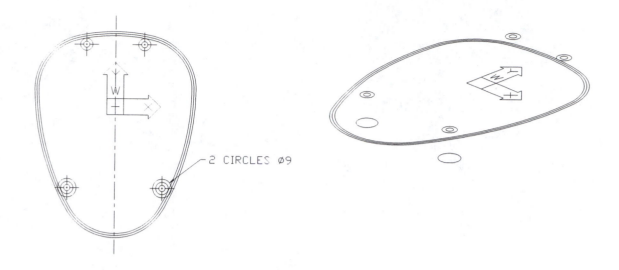

Figure 4.10 Two circles created and moved

Run the UCS command to set the UCS origin to (0,0,6).

[UCS]   **[Origin UCS]**

Command: **UCS**
Origin/ZAxis/3point/OBject/View/X/Y/Z/Prev/Restore/Save/Del/?/<World>: **OR**
Origin point: **0,0,6**

On the XY plane of the new UCS, create the wireframes as shown in Figure 4.11.

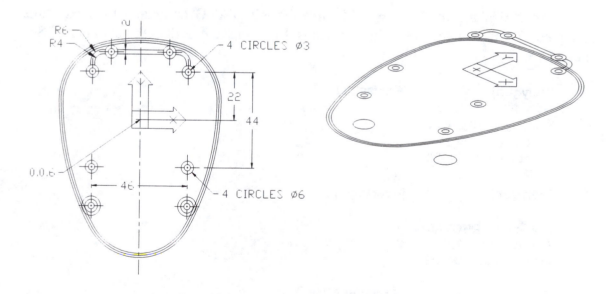

Figure 4.11 Wireframes drawn on the new UCS

Trim the circles A, B, C, and D. Then, convert the trimmed circles and those connected lines and arcs to a region. See Figure 4.12.

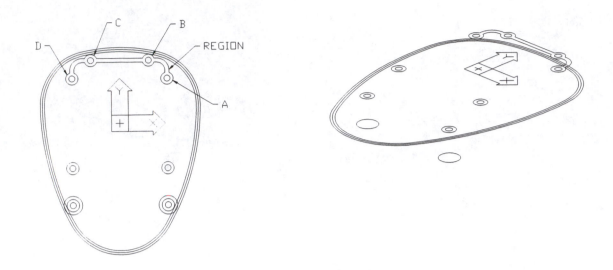

Figure 4.12 Trimmed circles and the connected lines and arcs converted to a region

Set the UCS origin to (*0,0,-12.5) by using the UCS command. The (*) prefix means that the coordinates are expressed as absolute world coordinate system values regardless of the current UCS orientation.

[UCS]   **[Origin UCS]**

Command: **UCS**
Origin/ZAxis/3point/OBject/View/X/Y/Z/Prev/Restore/Save/Del/?/<World>: **OR**
Origin point: ***0,0,-12.5**

On the XY plane of the new UCS, use the RECTANG command to make three rectangles. Then, convert them into regions by using the REGION command. See Figure 4.13.

[Polygon]   **[Rectangle]**

Command: **RECTANG**
First corner: **-22,20**
Other corner: **-24,-20**

[Polygon]   **[Rectangle]**

Command: **RECTANG**
First corner: **-22,-24**
Other corner: **-24,-38**

[Polygon]   **[Rectangle]**

Command: **RECTANG**
First corner: **-22,-42**
Other corner: **-24,-54**

[Polygon]   **[Region]**

```
Command: REGION
Select objects: [Select the three rectangles.]
Select objects: [Enter]
3 loops extracted
3 Regions created
```

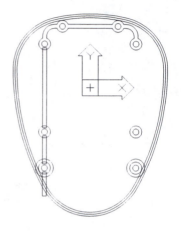
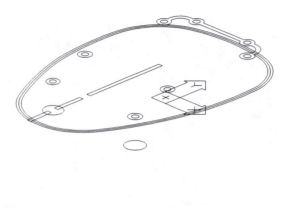

Figure 4.13 Three regions created

To make three more regions, run the MIRROR command to mirror the three rectangles about the Y axis. See Figure 4.14.

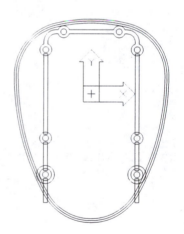
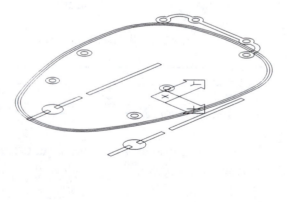

Figure 4.14 Three regions mirrored

Add two more regions and two circles. See Figure 4.15.

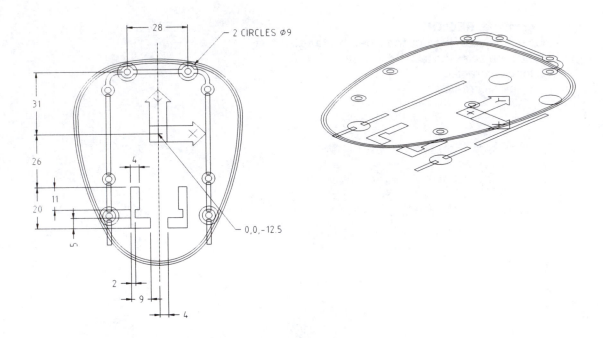

Figure 4.15 Two more regions and two circles created

Up to this point, the wireframes for the general structure of the lower casing are complete. Next, you will create the wireframes for the battery compartment.

Set the current layer to W_B and turn off the layer W_L.

<Data>   <Layers...>

Command: **DDLMODES**

Layer	
W_L	**Off**

Current layer: **W_B**

To proceed, create a number of entities as shown in Figure 4.16. After that, convert the entities to a region.

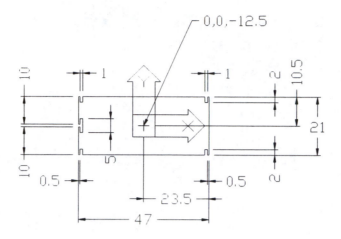

Figure 4.16 Wireframe for the battery compartment created

Create five more regions, as shown in Figure 4.17.

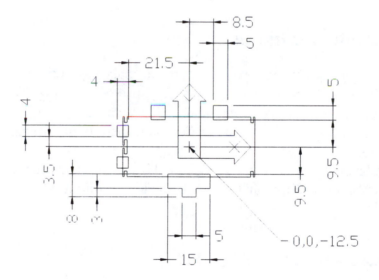

Figure 4.17 Five regions created

Make two more rectangular regions. See Figure 4.18.

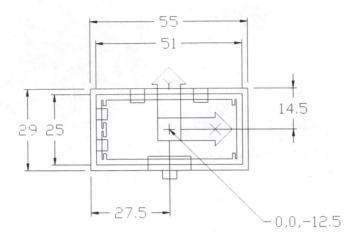

Figure 4.18 Two more rectangular regions created

All the wireframes for the lower casing are complete. Set the UCS back to WORLD by using the UCS command.

4.5   Solid Model for the Upper Casing

After creating all the wireframes, you will start to make the solid model for the upper casing.

As mentioned earlier, this is a thin shell model with internal webs and bosses. You need to build two solids, one for the outer skin as if there is no internal void and one to represent the void.

Turn off the layer W_B, turn on the layer W_U, and set the current layer to S_U. You will use the wireframes residing on layer W_U to build a solid model on layer S_U.

<Data>   <Layers...>

Command: **DDLMODES**

Layer	
W_B	**Off**
W_U	**On**

Current layer: **S_U**

Set to a single viewport. Then, use the VPOINT command to set the viewing position to rotate 345° in the XY plane and 35° from the XY plane. See Figure 4.19.

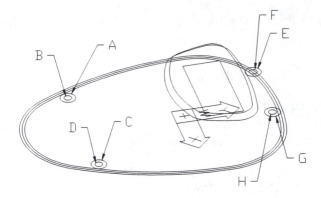

Figure 4.19 Wireframes for the upper casing

When you make the wireframes, you have to set the variable DELOBJ to 1 so the wireframes used to make the regions are deleted. This way, you will not duplicate the wireframes.

Now, you will use the wireframes for extruding or revolving. You should set the variable to 0 in order to keep the wireframes. Keeping the wireframes enables you to rebuild the model quickly if there is any mistake.

Command: **DELOBJ**
New value for DELOBJ: **0**

To begin, you will build the solid model that represents the inner void of the casing. Run the EXTRUDE command to extrude eight circles. They are the bosses for the upper casing. See Figure 4.20.

[Solids]   **[Extrude]**

Command: **EXTRUDE**
Select objects: **[Select A, B, C, D, E, F, G, and H (Figure 4.19).]**
Select objects: **[Enter]**
Path/<Height of Extrusion>: **20**
Extrusion taper angle: **0**

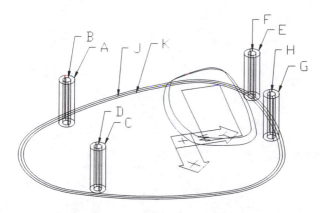

Figure 4.20 Eight circles extruded

After extrusion, run the SUBTRACT command to subtract the four smaller cylinders from the four larger cylinders.

[Modify]   **[Subtract]**

Command: **SUBTRACT**
Select solids and regions to subtract from...
Select objects: **[Select A, C, E, and G (Figure 4.20).]**
Select objects: **[Enter]**
Select objects: Select solids and regions to subtract...
Select objects: **[Select B, D, F, and H (Figure 4.20).]**
Select objects: **[Enter]**

Use the EXTRUDE command to extrude two wireframes, and use the SPHERE command to create a solid sphere. See Figure 4.21.

[Solids]   **[Extrude]**

Command: **EXTRUDE**
Select objects: **[Select J (Figure 4.20).]**
Select objects: **[Enter]**
Path/<Height of Extrusion>: **1**
Extrusion taper angle: **5**

[Solids]   **[Extrude]**

Command: **EXTRUDE**
Select objects: **[Select K (Figure 4.20).]**
Select objects: **[Enter]**
Path/<Height of Extrusion>: **20**
Extrusion taper angle: **5**

[Solids]   **[Sphere]**

Command: **SPHERE**
Center of sphere: ***0,5,-195**
Diameter/<Radius> of sphere: **210**

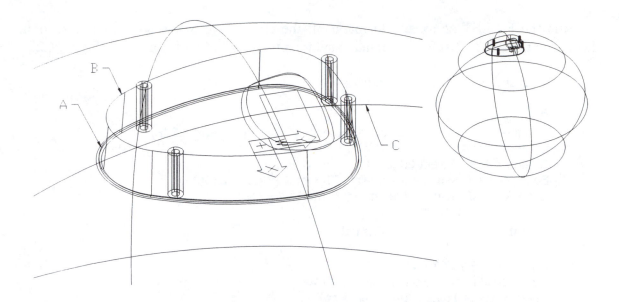

Figure 4.21 Extruded cylinders subtracted, the wireframes extruded, and a sphere created

Run the UNION command to unite the two newly extruded solids. Then, use the INTERSECT command on the united solid and the solid sphere. The result of intersection is the main body of the inner core. See Figure 4.22.

[Modify]   **[Union]**

Command: **UNION**
Select objects: **[Select A and B (Figure 4.21).]**
Select objects: **[Enter]**

[Modify]   **[Intersection]**

Command: **INTERSECT**
Select objects: **[Select B, and C (Figure 4.21).]**
Select objects: **[Enter]**

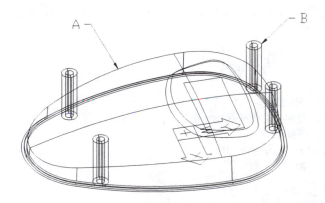

Figure 4.22 Two extruded solids united and intersected with the sphere

Run the FILLET command to round off the upper edge of the solid of intersection. Then, run the SUBTRACT command to subtract the four bosses from it. See Figure 4.23.

[Feature]   **[Fillet]**

Command: **FILLET**
Polyline/Radius/Trim/<Select first object>: **[Select A (Figure 4.22).]**
Enter radius: **4**
Chain/Radius/<Select edge>: **C**
Edge/Radius/<Select edge chain>: **[Select A (Figure 4.22).]**
Edge/Radius/<Select edge chain>: **[Enter]**

[Modify]   **[Subtract]**

Command: **SUBTRACT**
Select solids and regions to subtract from...
Select objects: **[Select the filleted solid.]**
Select objects: **[Enter]**
Select solids and regions to subtract...
Select objects: **[Select B (Figure 4.22).]**
Select objects: **[Enter]**

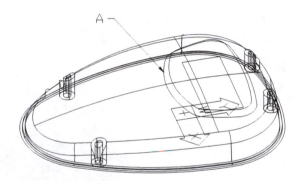

Figure 4.23 Edge filleted and four bosses subtracted

Run the EXTRUDE command to extrude a wireframe with a taper angle of -45°. See Figure 4.24.

[Solids]   **[Extrude]**

Command: **EXTRUDE**
Select objects: **[Select A (Figure 4.23).]**
Select objects: **[Enter]**
Path/<Height of Extrusion>: **20**
Extrusion taper angle: **-45**

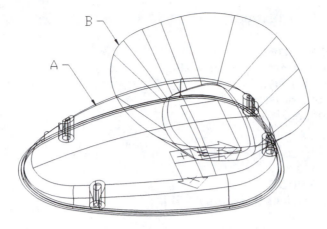

Figure 4.24 Wireframe extruded

Run the SUBTRACT command to subtract the extruded solid from the main body. See Figure 4.25.

[Modify]   [Subtract]

Command: **SUBTRACT**
Select solids and regions to subtract from...
Select objects: **[Select A (Figure 4.24).]**
Select objects: **[Enter]**
Select solids and regions to subtract...
Select objects: **[Select B (Figure 4.24).]**
Select objects: **[Enter]**

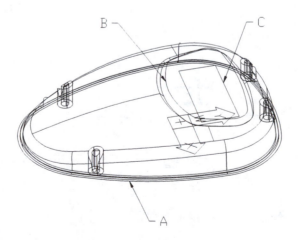

Figure 4.25 Completed solid core

The solid that represents the void of the upper casing is complete. To continue, you will work on the outer shell. Run the EXTRUDE command to extrude three wireframes, and use the SPHERE command to create a solid sphere. See Figure 4.26.

[Solids]   **[Extrude]**

Command: **EXTRUDE**
Select objects: **[Select A (Figure 4.25).]**
Select objects: **[Enter]**
Path/<Height of Extrusion>: **20**
Extrusion taper angle: **5**

[Solids]   **[Extrude]**

Command: **EXTRUDE**
Select objects: **[Select B (Figure 4.25).]**
Select objects: **[Enter]**
Path/<Height of Extrusion>: **20**
Extrusion taper angle: **-45**

[Solids]   **[Extrude]**

Command: **EXTRUDE**
Select objects: **[Select C (Figure 4.25).]**
Select objects: **[Enter]**
Path/<Height of Extrusion>: **20**
Extrusion taper angle: **0**

[Solids]   **[Sphere]**

Command: **SPHERE**
Center of sphere: ***0,5,-195**
Diameter/<Radius> of sphere: **212**

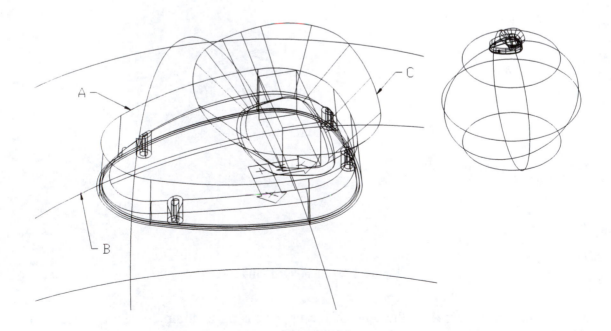

Figure 4.26 Three wireframes extruded and a sphere created

Run the INTERSECT command to intersect two solids and the SUBTRACT command to subtract another solid. See Figure 4.27.

[Modify]   **[Intersection]**

Command: **INTERSECT**
Select objects: **[Select A and B (Figure 4.26).]**
Select objects: **[Enter]**

[Modify]   **[Subtract]**

Command: **SUBTRACT**
Select solids and regions to subtract from...
Select objects: **[Select the newly created solid of intersection.]**
Select objects: **[Enter]**
Select solids and regions to subtract...
Select objects: **[Select C (Figure 4.26).]**
Select objects: **[Enter]**

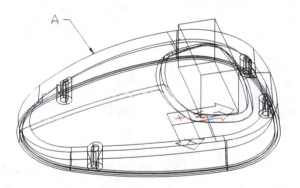

Figure 4.27 Solids intersected and subtracted

Run the FILLET command to fillet the upper edge of the outer skin. The fillet radius is the sum of the internal fillet radius and the thickness of the shell. See Figure 4.28.

[Feature]   **[Fillet]**

Command: **FILLET**
Polyline/Radius/Trim/<Select first object>: **[Select A (Figure 4.27).]**
Enter radius: **6**
Chain/Radius/<Select edge>: **C**
Edge/Radius/<Select edge chain>: **[Select A (Figure 4.27).]**
Edge/Radius/<Select edge chain>: **[Enter]**

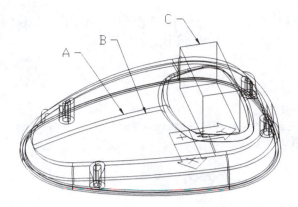

Figure 4.28 Outer skin completed

The solid that represents the outer skin is complete. Run the SUBTRACT command to subtract the inner core from the outer skin and to cut an opening in the casing.

[Modify]   **[Subtract]**

Command: **SUBTRACT**
Select solids and regions to subtract from...
Select objects: **[Select A (Figure 4.28).]**
Select objects: **[Enter]**
Select solids and regions to subtract...
Select objects: **[Select B and C (Figure 4.28).]**
Select objects: **[Enter]**

Turn off the layer W_U. The model for the upper casing is complete. See Figure 4.29.

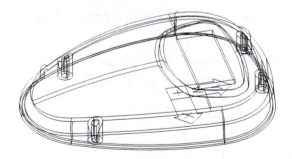

Figure 4.29 Completed upper casing

4.6   Solid Model for the Lower Casing

After making the model for the upper casing, you will continue to work on the model for the lower casing. The general approach is similar to that for the upper casing. You will create the solid skin, and then the solid core, and finally, you will subtract the solid core from the solid skin. Refer to Figure 4.2. You might find, in the assembly, that some part of the lower casing intrudes into the upper casing. The model of the lower casing is different from the upper casing in that its internal webs and bosses extend beyond the parting line of the thin shell. To tackle the problem, you can add a dummy solid box to the solid skin and another dummy solid box to the solid core. The size of the dummy boxes have to be large enough to encompass the portion that protrudes upwards.

Run the LAYER command to set the current layer to S_L1, turn off layer S_U, and turn on layer W_L. See Figure 4.30.

<Data>   <Layers...>

Command: **DDLMODES**

Layer	
S_U	Off
W_L	On

Current layer: **S_L1**

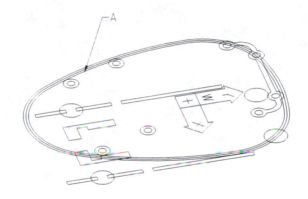

Figure 4.30 Wireframes for making the solid model of the lower casing

Use the MOVE command to translate the wireframe that represents the inner profile of the core a distance of 1 unit in the Z direction. Then, run the EXTRUDE command on it to create a solid of extrusion. After that, create a dummy solid box by using the BOX command. The dummy solid box and the extruded solid together form the main body of the solid core. See Figure 4.31.

[**Modify**]   [**Move**]

Command: **MOVE**
Select objects: **[Select A (Figure 4.30).]**
Select objects: **[Enter]**
Base point or displacement: **0,0,1**
Second point of displacement: **[Enter]**

[**Solids**]   [**Extrude**]

Command: **EXTRUDE**
Select objects: **P**
Select objects: **[Enter]**

Path/<Height of Extrusion>: **-11.5**
Extrusion taper angle: 5

[Solids]   **[Box]**   **[Corner]**

Command: **BOX**
Center/<Corner of box>: ***-50,-70,1**
Cube/Length/<other corner>: ***50,40,1**
Height: **20**

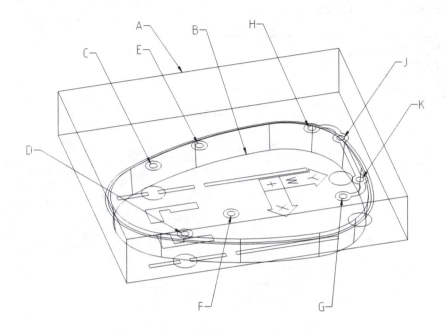

Figure 4.31 Wireframe moved and extruded, and a dummy solid box created

Use the UNION command to unite the dummy box and the solid of extrusion. Then, create eight solids of extrusion by using the EXTRUDE command. See Figure 4.32.

[Modify]   **[Union]**

Command: **UNION**
Select objects: **[Select A and B (Figure 4.31).]**
Select objects: **[Enter]**

[Solids]   **[Extrude]**

Command: **EXTRUDE**
Select objects: **[Select C, and D (Figure 4.31).]**
Select objects: **[Enter]**
Path/<Height of Extrusion>: **-12.5**
Extrusion taper angle: **0**

[Solids]   **[Extrude]**

Command: **EXTRUDE**
Select objects: **[Select E, F, G, and H (Figure 4.31).]**
Select objects: **[Enter]**
Path/<Height of Extrusion>: **-10**
Extrusion taper angle: **0**

[**Solids**]   [**Extrude**]

Command: **EXTRUDE**
Select objects: **[Select J, and K (Figure 4.31).]**
Select objects: **[Enter]**
Path/<Height of Extrusion>: -18.5
Extrusion taper angle: **0**

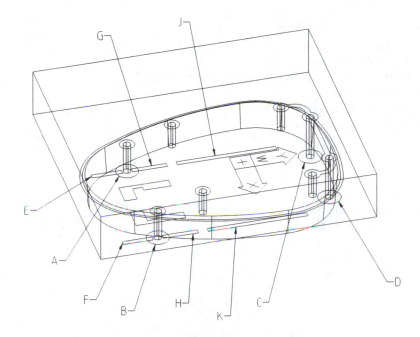

Figure 4.32 Dummy box united to the main core and the wireframes extruded

Create four solid cylinders by using the CYLINDER command. Then, extrude six rectangular regions and four circles by using the EXTRUDE command. See Figure 4.33.

[**Solids**]   [**Cylinder**]   [**Center**]

Command: **CYLINDER**
Elliptical/<center point>: **CEN** of **[Select A (Figure 4.32).]**
Diameter/<Radius>: **3**
Center of other end/<Height>: **3**

[**Solids**]   [**Cylinder**]   [**Center**]

Command: **CYLINDER**
Elliptical/<center point>: **CEN** of **[Select B (Figure 4.32).]**
Diameter/<Radius>: **3**
Center of other end/<Height>: **3**

[Solids]   **[Cylinder]**   **[Center]**

Command: **CYLINDER**
Elliptical/<center point>: **CEN** of **[Select C (Figure 4.32).]**
Diameter/<Radius>: **3**
Center of other end/<Height>: **3**

[Solids]   **[Cylinder]**   **[Center]**

Command: **CYLINDER**
Elliptical/<center point>: **CEN** of **[Select D (Figure 4.32).]**
Diameter/<Radius>: **3**
Center of other end/<Height>: **3**

[Solids]   **[Extrude]**

Command: **EXTRUDE**
Select objects: **[Select A, B, C, and D (Figure 4.32).]**
Select objects: **[Enter]**
Path/<Height of Extrusion>: **5**
Extrusion taper angle: **0**

[Solids]   **[Extrude]**

Command: **EXTRUDE**
Select objects: **[Select E, F, G, H, J, and K (Figure 4.32).]**
Select objects: **[Enter]**
Path/<Height of Extrusion>: **12.5**
Extrusion taper angle: **0**

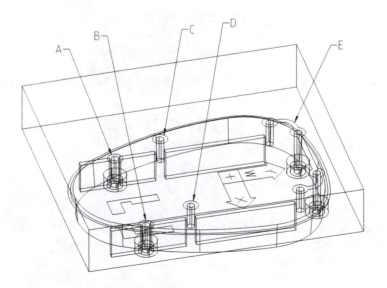

Figure 4.33 More solids created

Repeat the EXTRUDE command to make the bosses and webs. See Figure 4.34.

[Solids]　　　　　　**[Extrude]**

Command: **EXTRUDE**
Select objects: **[Select A and B (Figure 4.33).]**
Select objects: **[Enter]**
Path/<Height of Extrusion>: **-7.5**
Extrusion taper angle: **0**

[Solids]　　　　　　**[Extrude]**

Command: **EXTRUDE**
Select objects: **[Select C, D, and E (Figure 4.33).]**
Select objects: **[Enter]**
Path/<Height of Extrusion>: **-18.5**
Extrusion taper angle: **0**

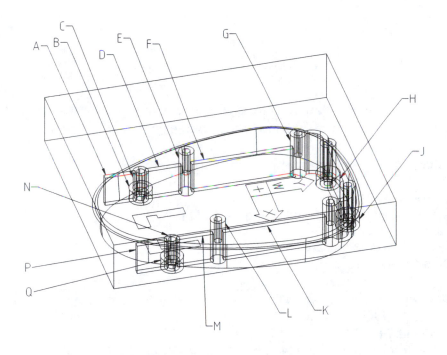

Figure 4.34 Bosses and webs created

Run the UNION command to unite the bosses and webs. See Figure 4.35.

[Modify]　　　　　　**[Union]**

Command: **UNION**
Select objects: **[Select A, B, C, D, E, F, G, H, J, K, L, M, N, P, and Q (Figure 4.34).]**
Select objects: **[Enter]**

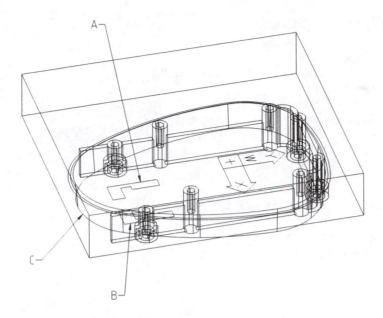

Figure 4.35 Bosses and webs united

Run the EXTRUDE command to create two solids of extrusion. Then, run the FILLET command to round off the lower edge of the core. See Figure 4.36.

[Solids]   **[Extrude]**

Command: **EXTRUDE**
Select objects: **[Select A and B (Figure 4.35).]**
Select objects: **[Enter]**
Path/<Height of Extrusion>: **21.5**
Extrusion taper angle: **0**

[Feature]   **[Fillet]**

Command: **FILLET**
Polyline/Radius/Trim/<Select first object>: **[Select C (Figure 4.35).]**
Enter radius: **4**
Chain/Radius/<Select edge>: **C**
Edge/Radius/<Select edge chain>: **[Select C (Figure 4.35).]**
Edge/Radius/<Select edge chain>: **[Enter]**

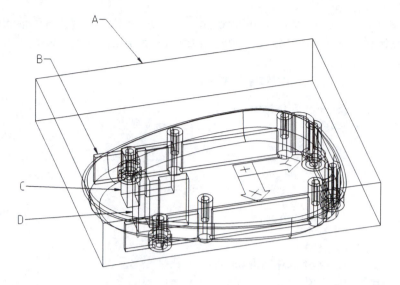

Figure 4.36 Two extruded solids created and the lower edge filleted

Use the SUBTRACT command to subtract the two newly extruded solids and the bosses and webs from the solid core. The solid that represents the core of the lower casing is complete. See Figure 4.37.

[Modify]   **[Subtract]**

Command: **SUBTRACT**
Select solids and regions to subtract from...
Select objects: **[Select A (Figure 4.36).]**
Select objects: **[Enter]**
Select solids and regions to subtract...
Select objects: **[Select B, C, and D (Figure 4.36).]**
Select objects: **[Enter]**

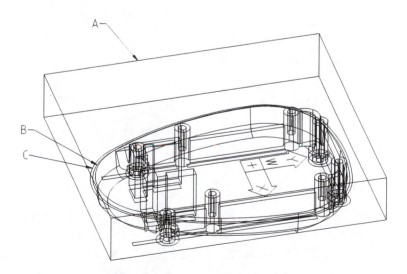

Figure 4.37 Solid that represents the inner core completed

Change the layer property of the core solid to layer S_L2. Then, use the EXTRUDE command to extrude two profiles for making the solid of the outer skin.

<Edit>   **<Properties...>**

Select objects: **[Select A (Figure 4.37).]**
Select objects: **[Enter]**

[Properties
Layer...   **S_L2**
OK   ]

[Solids]   **[Extrude]**

Command: **EXTRUDE**
Select objects: **[Select B (Figure 4.37).]**
Select objects: **[Enter]**
Path/<Height of Extrusion>: **1**
Extrusion taper angle: **5**

[Solids]   **[Extrude]**

Command: **EXTRUDE**
Select objects: **[Select C (Figure 4.37).]**
Select objects: **[Enter]**
Path/<Height of Extrusion>: **-12.5**
Extrusion taper angle: **5**

For the time being, the solid core and the wireframes are not needed. Turn off the layers S_L2 and W_L. See Figure 4.38.

<Data>   **<Layers...>**

Command: **DDLMODES**

Layer	
S_L2	Off
W_L	Off

Current layer: **S_L1**

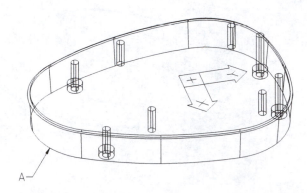

Figure 4.38 Two extruded solids created and the layer S_L2 turned off

Use the FILLET command to round off the lower edge of the outer skin. Then, run the BOX command to create a dummy box. This dummy box has to be smaller or equal to the dummy box for the inner core. See Figure 4.39.

[Feature]   **[Fillet]**

Command: **FILLET**
Polyline/Radius/Trim/<Select first object>: **[Select A (Figure 4.38).]**
Enter radius: **6**
Chain/Radius/<Select edge>: **C**
Edge/Radius/<Select edge chain>: **[Select A (Figure 4.38).]**
Edge/Radius/<Select edge chain>: **[Enter]**

[Solids]   **[Box]**   **[Corner]**

Command: **BOX**
Center/<Corner of box>: ***-48,-68,1**
Cube/Length/<other corner>: ***48,38,1**
Height: **20**

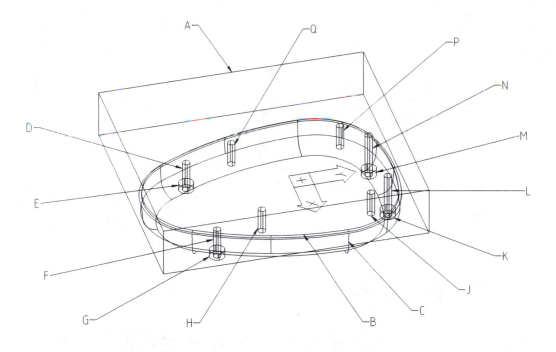

Figure 4.39 Outer skin filleted and the dummy box created

Run the UNION command to unite the two extruded profiles and the dummy box. Then, use the SUBTRACT command to subtract the holes and counterbores. After that, turn on the layer S_L2. See Figure 4.40.

[Modify]   **[Union]**

Command: **UNION**
Select objects: **[Select A, B, and C (Figure 4.39).]**
Select objects: **[Enter]**

[Modify]   **[Subtract]**

Command: **SUBTRACT**
Select solids and regions to subtract from...
Select objects: **[Select A (Figure 4.39).]**
Select objects: **[Enter]**
Select solids and regions to subtract...
Select objects: **[Select D, E, F, G, H, J, K, L, M, N, P, and Q (Figure 4.39).]**
Select objects: **[Enter]**

<Data>   **<Layers...>**

Command: **DDLMODES**

Layer	
S_L2	**On**

Current layer: **S_L1**

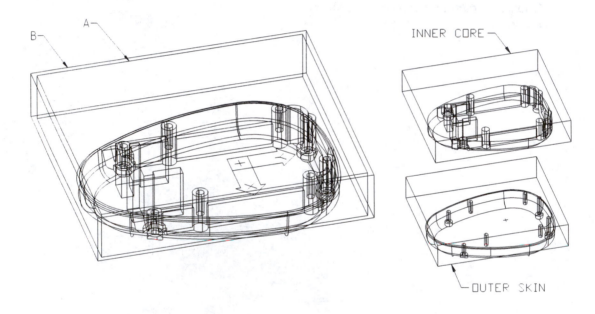

Figure 4.40 Outer skin completed and the inner core displayed

To complete the main body of the lower casing, run the SUBTRACT command to carve out the solid core from the solid outer skin. After carving, turn on the layer W_B. You will proceed to make the battery compartment. See Figure 4.41.

[Modify]   **[Subtract]**

Command: **SUBTRACT**
Select solids and regions to subtract from...
Select objects: **[Select A (Figure 4.40).]**
Select objects: **[Enter]**
Select solids and regions to subtract...
Select objects: **[Select B (Figure 4.40).]**
Select objects: **[Enter]**

<Data>   <Layers...>

Command: **DDLMODES**

Layer	
W_B	**On**

Current layer: **S_L1**

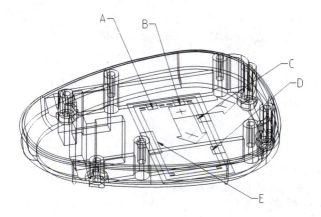

Figure 4.41 Inner solid core subtracted from the outer solid skin

At this point, the main part of the lower casing is complete. A battery compartment will finish the model.

Run the EXTRUDE command to create five extruded solids to use as openings in the battery compartment. See Figure 4.42.

[**Solids**]   [**Extrude**]

Command: **EXTRUDE**
Select objects: **[Select A, B, C, and D (Figure 4.41).]**
Select objects: **[Enter]**
Path/<Height of Extrusion>: **5**
Extrusion taper angle: **0**

[**Solids**]   [**Extrude**]

Command: **EXTRUDE**
Select objects: **[Select E (Figure 4.41).]**
Select objects: **[Enter]**
Path/<Height of Extrusion>: **10**
Extrusion taper angle: **0**

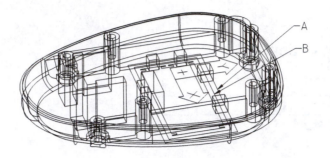

Figure 4.42 Five extruded solids created

Repeat the EXTRUDE command to make two extruded solids for adding to the lower casing to become the battery box. See Figure 4.43.

[Solids]   **[Extrude]**

Command: **EXTRUDE**
Select objects: **[Select A (Figure 4.42).]**
Select objects: **[Enter]**
Path/<Height of Extrusion>: **4**
Extrusion taper angle: **0**

[Solids]   **[Extrude]**

Command: **EXTRUDE**
Select objects: **[Select B (Figure 4.42).]**
Select objects: **[Enter]**
Path/<Height of Extrusion>: **14.5**
Extrusion taper angle: **0**

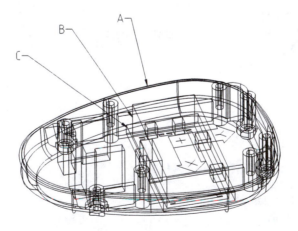

Figure 4.43 Two more wireframes extruded

Run the UNION command to add the last two solids to the lower casing of the main body. See Figure 4.44.

[Modify]   **[Union]**

Command: **UNION**
Select objects: **[Select A, B, and C (Figure 4.43).]**
Select objects: **[Enter]**

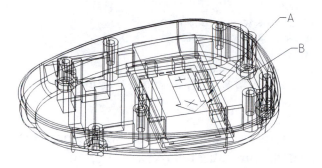

Figure 4.44 Solids united

Run the EXTRUDE command to extrude two solids for carving a shell for the battery case. See Figure 4.45.

[Solids]   **[Extrude]**

Command: **EXTRUDE**
Select objects: **[Select A (Figure 4.44).]**
Select objects: **[Enter]**
Path/<Height of Extrusion>: **2**
Extrusion taper angle: **0**

[Solids]   **[Extrude]**

Command: **EXTRUDE**
Select objects: **[Select B (Figure 4.44).]**
Select objects: **[Enter]**
Path/<Height of Extrusion>: **12.5**
Extrusion taper angle: **0**

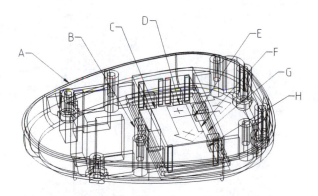

Figure 4.45 Two more wireframes extruded

To complete the lower casing, use the SUBTRACT command to carve out the unwanted solid.

[Modify]   **[Subtract]**

Command: **SUBTRACT**
Select solids and regions to subtract from...
Select objects: **[Select A (Figure 4.45).]**
Select objects: **[Enter]**
Select solids and regions to subtract...
Select objects: **[Select B, C, D, E, F, G, and H (Figure 4.45).]**
Select objects: **[Enter]**

The lower casing is now complete. The wireframes for making the battery compartment is not needed. Turn off the layer W_B. Figure 4.46 shows the front view, side view, top view, and isometric view of the lower casing.

<Data>   **<Layers...>**

Command: **DDLMODES**

Layer	
W_B	**Off**

Current layer: **S_L1**

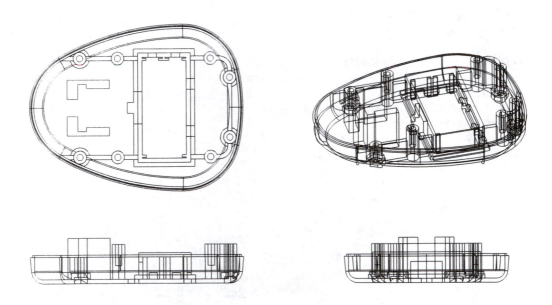

Figure 4.46 Completed lower casing

4.7   Utilities

After making the solid models for the upper and lower casings, you will learn how to check the interference between adjacent solid models, find out the mass properties of a solid model, create a 2D cross section across a solid model, and export the solid model in STL (Stereolithography) format for rapid prototyping and in 3DS (3D Studio) format for rendering and animation.

Interference checking is important to ensure that the solid parts mate together properly in an assembly. Evaluating the mass properties of a solid model is valuable to the designer analyzing the final product. Creating sections allows the designer to visualize the internal structure of the solid model more precisely. Regarding visualization, you can output a solid model to 3D Studio as meshed objects. Through the use of 3D Studio, you can create photo-realistic rendering and animation from a solid model.

Before you can check whether there is any interference between the upper casing and the lower casing, you have to turn on the layer S_U to display both the upper casing and the lower casing. See Figure 4.47.

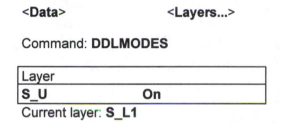

<Data>   **<Layers...>**

Command: **DDLMODES**

Layer	
S_U	**On**

Current layer: **S_L1**

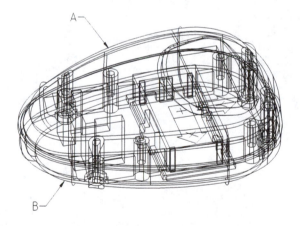

Figure 4.47 Upper casing and the lower casing put together

From the drawing shown, it is very hard to tell whether there is any interference between the upper and lower casings.

To be certain, you should use the INTERFERE command. Make a layer called UTY. Then, run the INTERFERE command on the upper and lower casings. If the command reports any interference, create the solid of interference.

<Data>   **<Layers...>**

Command: **DDLMODES**

Layer	
UTY	**cyan**

Current layer: **UTY**

[**Solids**]   [**Interfere**]

Command: **INTERFERE**
Select the first set of solids:
Select objects: **[Select A (Figure 4.47).]**
Select objects: **[Enter]**
Select the second set of solids:
Select objects: **[Select B (Figure 4.47).]**
Select objects: **[Enter]**
Comparing 1 solid against 1 solid.
Interfering solids (first set): 1
        (second set): 1
Interfering pairs: 1
Create interference solids ? <N>: **Y**

The volume of interference is represented as a solid. To see this solid clearly, you have to turn off the layers S_U and S_L1. See Figure 4.48.

<Data>   <Layers...>

Command: **DDLMODES**

Layer	
S_U	**Off**
S_L1	**Off**

Current layer: **UTY**

Figure 4.48 Solid of interference between the upper casing and the lower casing created

As you can see in Figure 4.48, the solid of interference is very small. Therefore, you can hardly notice it. Despite such small interference, it will prevent the upper and lower casings from fitting together properly. To ensure a perfect match between them, some solid from either casing has to be removed. If you study the design carefully, you will find that the solid of interference should be removed from the lower casing.

Use the GROUP command to put the solid of interference in an object group called INTER. Then, turn on the layer S_L1 by using the DDLMODES command. See Figure 4.49.

[Standard Toolbar]   **[Object Group]**

Command: **GROUP**

       [Group name: **INTER**
         New   ]

Select objects for grouping:
Select objects: **[Select A (Figure 4.48).]**
Select objects: **[Enter]**

       [OK   ]

<Data>   **<Layers...>**

Command: **DDLMODES**

Layer	
S_L1	**On**

Current layer: **UTY**

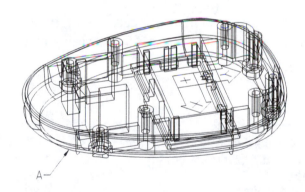

Figure 4.49   Lower casing and the solid of interference

To remove interference, use the SUBTRACT command to subtract the solid of interference from the lower casing. Then, turn on the layer S_U again. See Figure 4.50.

[Modify]   **[Subtract]**

Command: **SUBTRACT**
Select solids and regions to subtract from...
Select objects: **[Select A (Figure 4.49).]**
Select objects: **[Enter]**
Select solids and regions to subtract...
Select objects: **GROUP**
Enter group name: **INTER**
Select objects: **[Enter]**

<Data>   **<Layers...>**

Command: **DDLMODES**

Layer	
S_U	**On**

Current layer: **UTY**

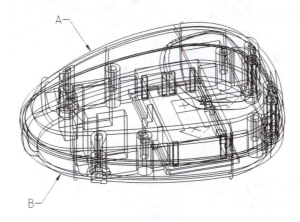

Figure 4.50 Upper casing and the modified lower casing

After subtracting the extra solid that causes interference, you can run the INTERFERE command again to ensure that the upper casing and the lower casing do not interfere with each other.

[Solids]   **[Interfere]**

Command: **INTERFERE**
Select the first set of solids:
Select objects: **[Select A (Figure 4.50).]**
Select objects: **[Enter]**
Select the second set of solids:
Select objects: **[Select B (Figure 4.50).]**
Select objects: **[Enter]**
Comparing 1 solid against 1 solid.
Solids do not interfere.

As expected, there is no interference between the upper and lower casings. Now, you can safely assemble them together.

Because a solid model consists of volume data as well as edge and surface data, you can determine the mass properties of a solid. Run the MASSPROP command. The default density is one.

[Object Properties]   **[Mass Properties]**

Command: **MASSPROP**
Select objects: **[Select B (Figure 4.50).]**
Select objects: **[Enter]**

```
------------ SOLIDS ------------
Mass:        29523.9455
Volume:       29523.9455
Bounding box:    X: -62.7393 -- 38.7992
        Y: -30.1672 -- 23.6664
        Z: -37.8488 -- 32.3096
Centroid:      X: -9.3315
        Y: -5.9167
        Z: -1.0895
Moments of inertia:  X: 14779002.1910
        Y: 33736037.2289
        Z: 30687059.1249
Products of inertia: XY: 3254018.4516
        YZ: -5478775.5495
        ZX: -1846072.8187
Radii of gyration:  X: 22.3736
        Y: 33.8034
        Z: 32.2397
-- Press RETURN for more --
Principal moments and X-Y-Z directions about centroid:
        I: 12993216.3572 along [0.9663 0.1516 -0.2082]
        J: 35126291.3792 along [-0.0044 0.8179 0.5753]
        K: 23803715.4459 along [0.2575 -0.5550 0.7910]

Write to a file ? <N>: N
```

In the process of product design, you might want to examine the internal structure of a solid. To aid visualizing, you can create a section by using the SECTION command. Run the command to create a cross section across the two solids.

[Solids]   **[Section]**

Command: **SECTION**
Select objects: **[Select A and B (Figure 4.50).]**
Select objects: **[Enter]**
Section plane by Object/Zaxis/View/XY/YZ/ZX/<3points>: **YZ**
Point on YZ plane: **0,0,0**

To see the cross sections clearly, turn off the layers S_U, and S_L1. See Figure 4.51.

<Data>   **<Layers...>**

Command: **DDLMODES**

Layer	
S_U	Off
S_L1	Off

Current layer: **UTY**

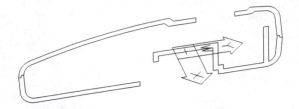

Figure 4.51 Section across the two solids

After seeing the cross section, set the current layer to S_U, turn on layer S_L1, and turn off layer UTY.

<Data>   <Layers...>

Command: **DDLMODES**

Layer	
S_L1	**On**
UTY	**Off**

Current layer: **S_U**

To output a solid model for making a rapid prototype, you can use the STLOUT command. Before doing that, you should adjust the resolution of the STL (Stereolithography) model by setting the FACETRES system variables. The value of this variable ranges from 0.01 to 10.0. A value of 10 gives the highest resolution.

Command: **FACETRES**
New value for FACETRES: **0.5**

Before you can output the file in STL format, you have to translate the solid model so that the entire model lies on the positive side of the X, Y, and Z axes.

Command: **STLOUT**
Select a single solid for STL output:
Select object: **[Select the lower casing.]**
Select object: **[Enter]**
Create a STL binary STL file<Y>: **Y**

A solid model contains data about the faces and volumes of an object. You can use the data of the surface to output photo-realistic rendering and animation by using 3D Studio. To convert the data for use in 3D Studio, run the 3DSOUT command.

Command: **3DSOUT**
Select objects: **[Select an object.]**
Select objects: **[Enter]**

4.8   The Battery Cover

After making the upper and lower casings under guidance, you will create the solid model for the lower casing on your own. Figure 4.52 is an orthographic drawing that shows the dimensions of this battery cover.

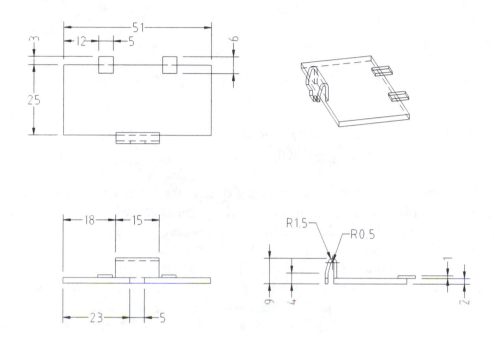

Figure 4.52 Battery cover dimensions

To build this solid model, you can continue to work on the drawing file that contains the upper casing and the lower casing, or you can start a new drawing. Either way, you should use the wireframes that you have already created for the battery compartment of the lower casing. See Figures 4.16, 4.17, and 4.18.

If you continue to work on the existing drawing, you should make a new layer for the new solid part, turn on the layer W_B, and turn off all other layers. If you prefer to start a new drawing, you can use the WBLOCK command to output the wireframes that reside on the layer W_B to a new drawing and then proceed with model making.

Using the wireframes that reside on the layer W_B, you can start to build the solid model. After making the battery cover, you should check for any interference.

If you start a new drawing, you will have to bring the battery cover to the lower casing by using the INSERT command. After insertion, the object shown in the drawing is an instance of an inserted block. Before you can process it like any other solid, you need to run the EXPLODE command on it. When you use this command on an inserted solid, you must be very careful not to overexplode. The process of exploding an inserted solid works as follows: When you explode the inserted solid once, you create a copy of the inserted solid at the insertion point. If you explode it twice or more, you will break it down into separate line and arc entities. Of course, you can UNDO the EXPLODE command. If you are not sure whether you have exploded it or not, you can use the LIST command to check the data type. It should be a solid.

To reduce the memory size after exploding the inserted solid, you have to use the PURGE command to remove the unreferenced inserted block.

After inserting, exploding, and purging, you can use the INTERFERE command to check if there is any interference between the battery cover and the lower casing.

4.9 Summary

In this chapter, you practiced the following AutoCAD commands related to 3D solid modeling.

INTERFERE	MASSPROP	SECTION
STLOUT	3DSOUT	

For a brief explanation of these commands, refer to the appendix of this book.

Through the project depicted in this chapter, you have gained an appreciation of how to create a thin shell solid model with internal webs and bosses. In addition, you have learned how to check two sets of solids for interference, how to evaluate the mass properties of a solid model, how to check the minute details of a model by creating a section, and how to output a solid model in Stereolithography format for making a rapid prototype and in 3D Studio format for generating photo-realistic rendering and animation.

In the next chapter, you will learn how to prepare a document from a solid model.

4.10 Exercise

To enhance your knowledge of making a thin shell solid model, you will produce a model of the casing of a water filtration unit. Figure 4.53 shows the engineering drawing of this model.

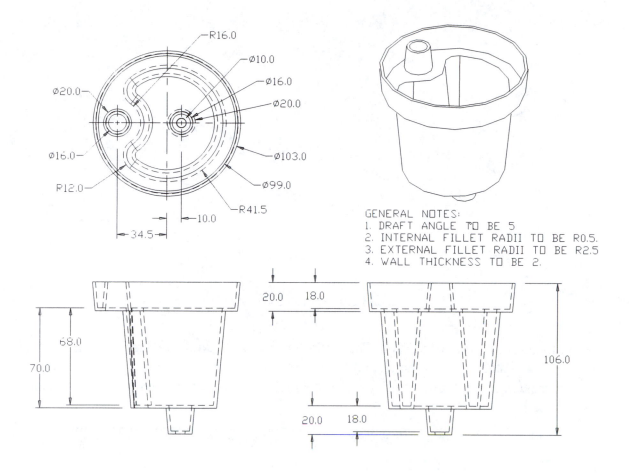

Figure 4.53 Casing for the water filtration unit

Start a new drawing. Make two layers called WIRE and SOLID, and set the layer WIRE as the current layer. You will build the wireframes on this layer.

On the layer WIRE, create a number of circles and arcs. Change them to nine regions or polylines. Then, move the wires A and B a distance of (0,0,-20), the wires C, D, and E a distance of (0,0,-86), and the wires F and G a distance of (0,0,-18). See Figure 4.54.

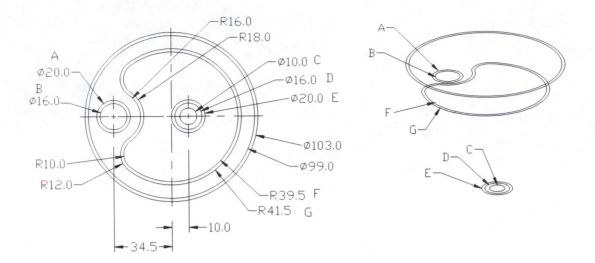

Figure 4.54 Set of circles and arcs drawn

Set the current layer to SOLID. Then, extrude the Ø103 circle a distance of -20 units, the wire G a distance of -70 units, and the wire E a distance of -20 units. The draft angle of all extrusion is 5°. Unite the three extruded solids. This is the outer skin of the solid model. See Figure 4.55.

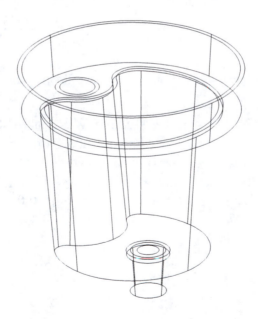

Figure 4.55 Outer solid skin created

To create the inner core of the solid, extrude the Ø99 circle a distance of -18 units, the wire F a distance of -68 units, and the wire D a distance of -18 units. The draft angle of all extrusion is 5°. Unite the three extruded solids. This is the core solid. Subtract it from the outer skin. See Figure 4.56.

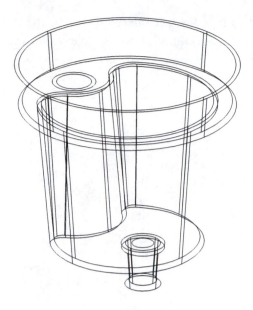

Figure 4.56 Inner solid core created

To continue, extrude the wires A and B a distance of 20 units and the wire C a distance of -20 units. The draft angle of all extrusion is 5°. See Figure 4.57.

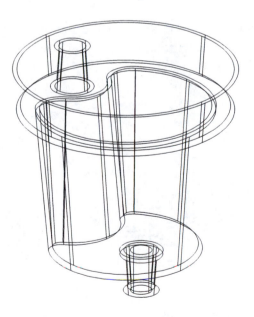

Figure 4.57 Three more wires extruded

To complete the model, unite the extruded solid A to the main body, and subtract the extruded solids B and C. To see the model more clearly, set the ISOLINES variable to 0, and the variable DISPSILH to 1. See Figure 4.58.

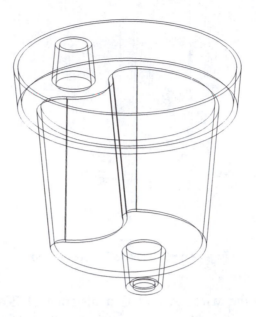

Figure 4.58 Completed water filtration model

To further modify the model, you can fillet the internal edges with a radius of R0.5 and the external edges R2.5. See Figure 4.59.

Figure 4.59 Water filtration model filleted

Chapter 5
Documentation

In addition to outputting a solid model in electronic data format for onward processing, you might need to prepare an engineering document from it. To prepare a document, you can use the paper-space environment of AutoCAD. Recall that there are two working environments in AutoCAD: the model-space environment and the paper-space environment. The model-space environment, the default working environment, is where you create the solid model. The paper-space environment is the environment where you can prepare a document from the solid model. To switch from the model-space environment to the paper-space environment, you should set the system variable TILEMODE to zero. To return to the model-space environment, simply set the variable to one.

In this chapter, you will prepare a document for the solid model of the hydraulic valve that you built in Chapter 3.

Document preparation involves two major tasks: adding a title block in paper space and creating floating viewports in paper space.

5.1   Title Block

Before you start to prepare a document from a solid model, you should already have a drawing that contains the title block. Typically, a title block should include the following information:
1. Title of the drawing
2. Scale of the drawing
3. Date of the drawing
4. Units of measurement
5. Name of the draftsperson
6. Name of the company
7. Drawing number
8. Number of sheets
9. Revision information

Figure 5.1 shows a typical title block. In making the title block, you might ask: What is the width of the margin between the four borders and the edges of the paper. The answer is: It depends on how wide your plotting device needs to be to clamp the paper while it prints. You can determine this by drawing a rectangle. Then, plot the rectangle with your plotting device using the "Fit to size" option. After plotting, measure the lengths of the lines printed. These lengths are the width and height of the rectangle that you should use to create a formal title block.

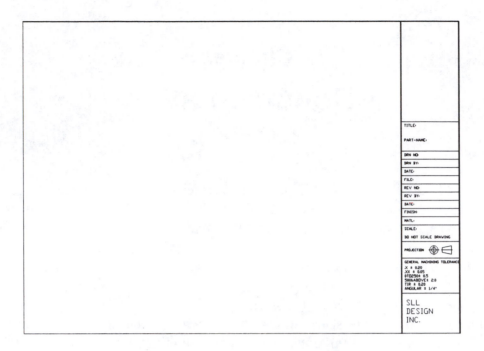

Figure 5.1 Typical drawing title block

5.2   Adding a Title Block

Open the drawing of the hydraulic valve. By default, this model was created in the model-space environment. See Figure 5.2.

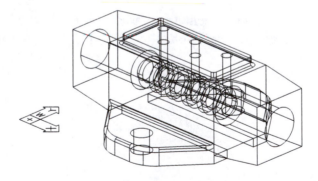

Figure 5.2 Solid model

To begin making an engineering document, switch the display to the paper space environment by setting the system variable TILEMODE to zero.

Command: **TILEMODE**
New value for TILEMODE: **0**

<View>   <Paper Space>

After setting the system variable TILEMODE to zero, the solid model in the model space disappears temporarily. The model is not displayed on the screen because you have switched to paper-space environment.

Now, there should be nothing on your screen. The default screen display size is the size defined by the LIMITS command. This size depends on the limits of the prototype drawing that you used to create the drawing. This display size might not be suitable for preparing a drawing. Before you can create floating viewports in the paper space, you must either zoom to a known size or place a title block of known size on the drawing.

Prior to inserting a drawing title block, make a new layer called TITLE. You will put the title block on this layer.

<Data>   **<Layers...>**

Command: **DDLMODES**

Layer
TITLE

Current layer: **TITLE**

To insert the title block on your drawing, run the INSERT command. See Figure 5.3.

Command: **INSERT**
Block name (or ?): **[Specify the filename of the title block.]**
Insertion point: **0,0**
X scale factor <1> / Corner / XYZ: **1**
Y scale factor (default=X): **1**
Rotation angle: **0**

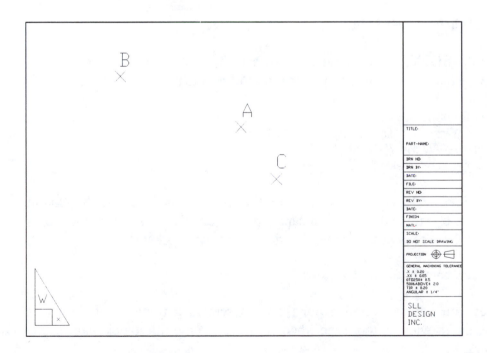

Figure 5.3 Title block inserted in paper space

5.3   Creating Floating Viewports

With a proper title block of known size in position, you can make the floating viewports. To do so, you can use the SOLVIEW command. This command is tailor-made for solid models. In addition to making the floating viewports like the MVIEW command does, the SOLVIEW command:

1. Creates a new layer called VPORTS for holding the viewport entities.
2. Prompts you to input a viewport name for each floating viewport. Then, it creates three additional layers. The name of the first layer uses the viewport name with the suffix "-DIM," the second with "-HID," and the third with "-VIS." The first layer, -DIM, is reserved for placing the dimension entities. The layers -HID and -VIS are used for placing the hidden lines and visible lines that are projected from the solid model by the SOLVIEW command. You should not put anything on these two layers. If you have already loaded the linetype HIDDEN, the layer -HID will use this linetype.
3. Allows you to create an orthographic viewport that aligns with a chosen drawing viewport.
4. Enables you to prepare a viewport for auxiliary view.
5. Allows you to prepare a viewport for projecting a sectional view by using the SOLDRAW command. If you specify a sectional view, a fourth layer will be created for that viewport, with the suffix "-HAT".
6. Freezes the additional layers in all other floating viewports so that they are visible only in the specific viewport.
7. Saves a display view using the name of the viewport. You can return to this saved view anytime after zooming and panning.

Use the LINETYPE command to load the linetype HIDDEN to the drawing file.

    Command: **LINETYPE**

Run the SOLVIEW command. To start with, use the UCS option and set the view to WORLD. This will produce a view that resembles the plan view of the WCS.

    Command: **SOLVIEW**
    Ucs/Ortho/Auxiliary/Section/<eXit>: **U**
    Named/World/?/<Current>: **W**
    Enter view scale: **1**

The view scale is a zoom scale of the objects displayed in the floating viewport relative to the entities displayed in paper space. Tentatively, set it to one. After that, select a point on the screen as the center position of the viewport.

    View center: **[Select A (Figure 5.3).]**

After you have selected a point on the screen, a preview of the objects appears. Because the preview is too large when compared to the title block, cancel the SOLVIEW command.

    View center: ***Cancel***

Run the SOLVIEW command again. This time, specify a view scale of 0.5.

```
Command: SOLVIEW
Ucs/Ortho/Auxiliary/Section/<eXit>: U
Named/World/?/<Current>: W
Enter view scale: 0.5
View center: [Select A (Figure 5.3).]
View center: [Enter]
Clip first corner: [Select B (Figure 5.3.).]
Clip other corner: [Select C (Figure 5.3).]
View name: TOP
```

At this point, a top view with a relative zoom scale of 0.5 appears. See Figure 5.4. If you run the 'DDLMODES command transparently, you will find that the four layers VPORTS, TOP-DIM, TOP-HID, and TOP-VIS are created.

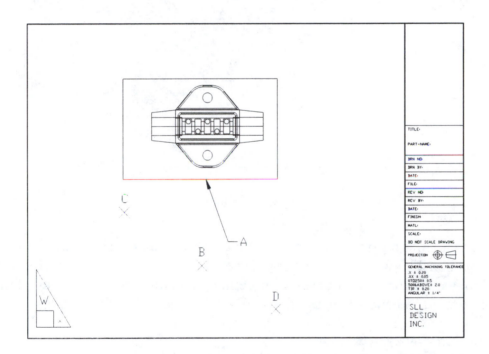

Figure 5.4 Floating viewport for the top view created

Continue the SOLVIEW command to create a front view, which is an orthographic view. To make an orthographic view, you have to choose the O option. An orthographic view needs to be projected from an existing view; therefore, you have to select an edge of the top view to specify from which direction you would like to project. Select the lower edge of the top view, and then specify the location and the size of the new viewport. See Figure 5.5.

```
Ucs/Ortho/Auxiliary/Section/<eXit>: O
Pick side of viewport to project: [Select A (Figure 5.4).]
View center: [Select B (Figure 5.4).]
View center: [Enter]
Clip first corner: [Select C (Figure 5.4).]
```

Clip other corner: **[Select D (Figure 5.4).]**
View name: **FRONT**

Again, you will have created three additional layers, FRONT-DIM, FRONT-HID, and FRONT-VIS.

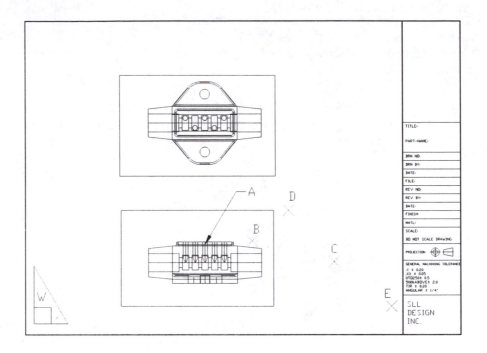

Figure 5.5 Floating viewport for the front view created

Next, you will prepare a viewport for projecting a sectional side view from the front view. A sectional view requires a cutting plane; therefore, you need to specify two points on the front view for this purpose. After defining a cutting plane, you have to indicate which side to project and where the viewport will be placed. See Figure 5.6.

Ucs/Ortho/Auxiliary/Section/<eXit>: **S**
Cutting Plane's 1st point: **CEN** of **[Select A (Figure 5.5).]**
Cutting Plane's 2nd point: **@100<270**
Side to view from: **[Select B (Figure 5.5).]**
Enter view scale: **0.5**
View center: **[Select C (Figure 5.5).]**
View center: **[Enter]**
Clip first corner: **[Select D (Figure 5.5).]**
Clip other corner: **[Select E (Figure 5.5).]**
View name: **SECT**
Ucs/Ortho/Auxiliary/Section/<eXit>: **[Enter]**

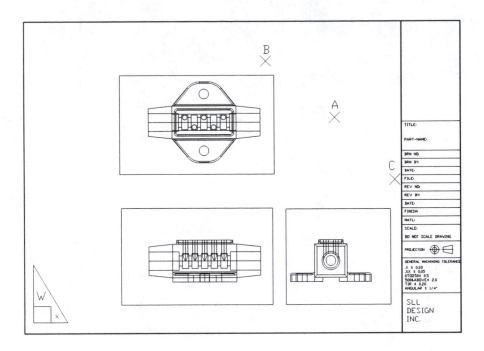

Figure 5.6 Floating viewport for the sectional side view created

Now, you have created three aligned floating viewports. You can regard them as view windows through which you see the solid model in three directions. The additional layers created by the SOLVIEW command are, at this moment, empty. Later, you will use the SOLDRAW command on these viewports to generate 2D orthographic views on the additional layers.

Before you do that, you will add two more viewports — an isometric view and an auxiliary view.

For the isometric view, run the SOLVIEW command to make a floating viewport that is independent of the three existing viewports. See Figure 5.7.

```
Command: SOLVIEW
Ucs/Ortho/Auxiliary/Section/<eXit>: U
Named/World/?/<Current>: W
Enter view scale: 0.5
View center: [Select A (Figure 5.6).]
View center: [Enter]
Clip first corner: [Select B (Figure 5.6).]
Clip other corner: [Select C (Figure 5.6).]
View name: 3D
Ucs/Ortho/Auxiliary/Section/<eXit>: [Enter]
```

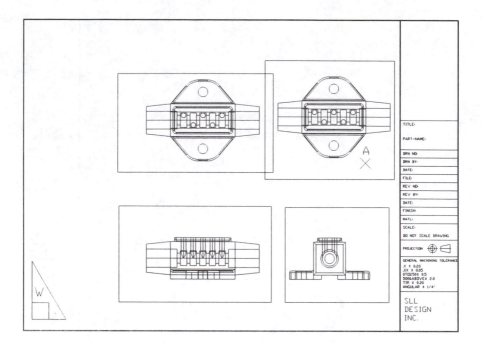

Figure 5.7 Fourth floating viewport created

For the time being, the fourth viewport displays the top view of the model. Before you can change it to a 3D view, you have to enter into the model space in the floating viewport. Run the MSPACE command.

<View>   <Floating Model Space>

Command: **MSPACE**

Select A (Figure 5.7) within the fourth viewport to choose it as the current viewport. Then, run the VPOINT command to display a 3D view.

<View>   <3D Viewpoint>   <Vector>

Command: **VPOINT**
Rotate/<View point>: **R**
Enter angle in XY plane from X axis: **315**
Enter angle from XY plane: **25**

Although the VPOINT command creates a display that occupies the entire area of the floating viewport, it does not allow you to specify a particular zoom scale relative to the paper space.

To zoom the view properly, use the ZOOM command to adjust the relative zoom scale to 0.5XP. See Figure 5.8.

[Zoom]   [Zoom Scale]

Command: **ZOOM**
All/Center/Dynamic/Extents/Left/Previous/Vmax/Window/<Scale(X/XP)>: **0.5XP**

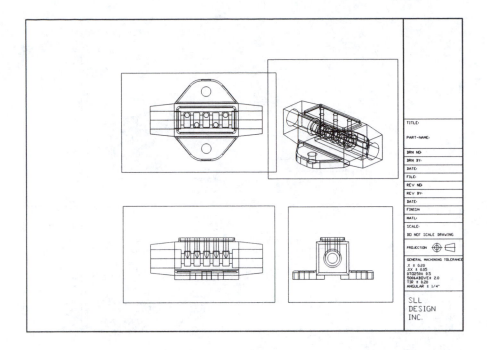

Figure 5.8 Fourth floating viewport changed to an isometric view

In addition to making an orthographic 3D view, you can prepare a perspective view. Run the DVIEW command. Set the camera to 35° from the XY plane, and -65° from the X axis in the XY plane. Then, choose the DISTANCE option to set camera distance and to turn on the perspective mode. See Figure 5.9.

```
<View>                          <3D Dynamic View>

Command: DVIEW
Select objects: [Select the solid model.]
Select objects: [Enter]
CAmera/TArget/Distance/POints/PAn/Zoom/TWist/CLip/Hide/Off/Undo/<eXit>: CA
Toggle angle in/Enter angle from XY plane: 35
Toggle angle from/Enter angle in XY plane from X axis: -65
CAmera/TArget/Distance/POints/PAn/Zoom/TWist/CLip/Hide/Off/Undo/<eXit>: D
New camera/target distance: 350
CAmera/TArget/Distance/POints/PAn/Zoom/TWist/CLip/Hide/Off/Undo/<eXit>: [Enter]
```

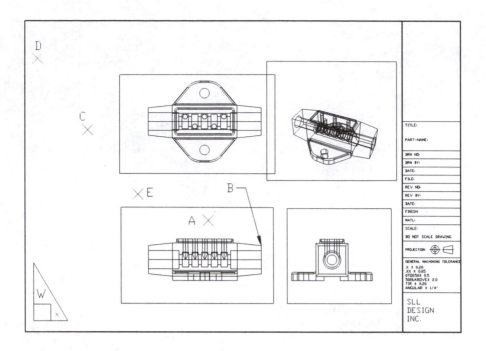

Figure 5.9 Fourth floating viewport changed to a perspective view

As mentioned earlier, the SOLVIEW command saves the display view using the name of the floating viewport. To check these saved views, you can use the VIEW command.

```
Command: VIEW
?/Delete/Restore/Save/Window: ?
View(s) to list <*>: [Enter]
Saved views:
View name              Space
3D                     M
FRONT                  M
SECT                   M
TOP                    M
```

As you can see, each of the four saved display views remembers the corresponding floating viewport. Because you have changed the display of the 3D viewport to an isometric view and then to a perspective view, you have to overwrite the saved view 3D by running the VIEW command again. Before you run this command, make sure that the perspective view is the current viewport.

```
Command: VIEW
?/Delete/Restore/Save/Window: SAVE
View name to save: 3D
```

Return to the paper space by using the PSPACE command.

```
<View>                      <Paper Space>

Command: PSPACE
```

After returning to the paper space, run the SOLVIEW command again to produce an auxiliary viewport. Quite similar to making a viewport for a sectional view, you have to specify a plane by defining two points on a chosen viewport. See Figure 5.10.

```
Command: SOLVIEW
Ucs/Ortho/Auxiliary/Section/<eXit>: A
Inclined Plane's 1st point:    [Select A (Figure 5.9) to select the front viewport.]
              END of [Select B (Figure 5.9).]
Inclined Plane's 2nd point: @100<225
Side to view from: [Select A (Figure 5.9).]
View center: [Select C (Figure 5.9).]
View center: [Enter]
Clip first corner: [Select D (Figure 5.9).]
Clip other corner: [Select E (Figure 5.9).]
View name: AUX
Ucs/Ortho/Auxiliary/Section/<eXit>: [Enter]
```

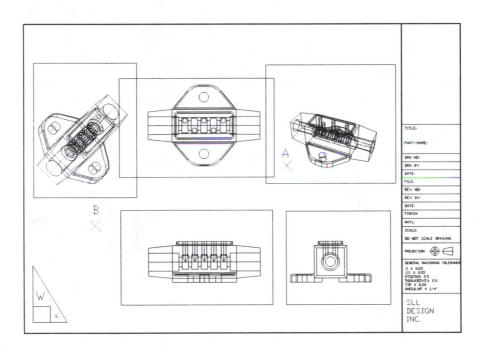

Figure 5.10 Floating viewport for the auxiliary view created

At this moment, the floating viewports are simply viewing windows through which you can see the 3D solid model in several directions. To project 2D views from the 3D solid model, you have to run the SOLDRAW command, which does the following:

1. Projects the solid model onto a 2D plane to generate visible outlines. These lines reside on the layer "-VIS".
2. Projects the solid model onto a 2D plane to generate hidden lines. These lines reside on the layer "-HID".
3. Freezes the layer that holds the solid model in all the floating viewports.
4. Projects a sectional view onto a 2D plane if the viewport is prepared as a sectional viewport. The hatching lines reside on the layer "-HAT".

Command: **SOLDRAW**
Select viewports to draw:
Select objects: **[Select A (Figure 5.10).]**
Other corner: **[Select B (Figure 5.10).]**
Select objects: **[Enter]**

By default, AutoCAD linetypes are defined in English units. Because this book uses metric units, you need to change the linetype scale to 25. Run the LTSCALE command. See Figure 5.11.

<Options>   **<Linetypes>**   **<Global Linetype Scale>**

Command: **LTSCALE**
New scale factor: **25**

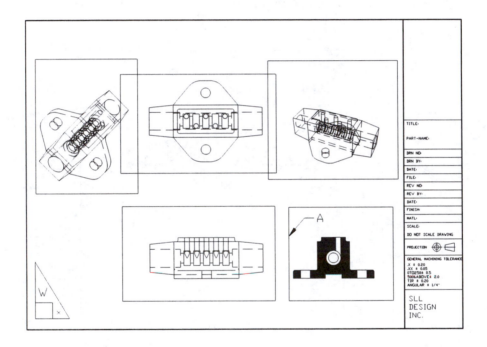

Figure 5.11 Visible lines and hidden lines projected

As you can see in Figure 5.11, the hatching lines are too dense. To change the hatching density, you have to manipulate the variable HPSCALE, and then run the SOLDRAW command again.

When you use the SOLDRAW command on a viewport, it will first erase all the entities on the layers "-HID," "-VIS," and "-HAT." Then, it generates the visible lines, hidden lines, and hatching lines again. Consequently, you should not use these layers for any other purpose because everything you put there might be erased.

Set the hatching scale and generate the sectional view again.

Command: **HPSCALE**
New value for HPSCALE: **70**

Command: **SOLDRAW**

```
Select viewports to draw:
Select objects: [Select A (Figure 5.11).]
Select objects: [Enter]
```

If you watch carefully, you will notice that the entities are erased and then regenerated on this viewport.

To complete the drawing, you should freeze the layer VPORTS for two reasons: (1) The borders of the floating viewports reside on this layer, so you have to freeze this layer to hide the viewport borders, and (2) The viewports had been aligned properly by the SOLVIEW command, so freezing this layer prevents you from accidentally jeopardizing the alignment of the floating viewports.

Freeze the layer VPORTS. See Figure 5.12. The engineering drawing views are complete.

<Data>   **<Layers...>**

Command: **DDLMODES**

Layer	
VPORTS	**Freeze**

Current layer: **TITLE**

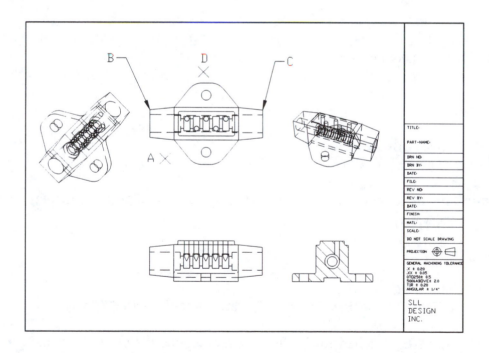

Figure 5.12 Sectional view erased and regenerated

To summarize, the SOLVIEW command and the SOLDRAW command work together to produce the engineering drawing views in the paper-space environment. The SOLVIEW command creates the floating viewports and the related layers, and saves the display views. The SOLDRAW command generates the 2D views on the floating viewports that are created by the SOLVIEW command. This command does not work on the floating viewports that are created by the MVIEW command. For such viewports, you can use the SOLPROF command. Both the SOLDRAW command and the SOLPROF command work similarly; one is used for viewports created by the SOLVIEW command and the other one is used for viewports created by the MVIEW command.

5.4 Dimensioning

It is common practice to include dimensions on a drawing. Each of the floating viewports has a layer suffixed by the letters "-DIM," which is reserved for dimensioning.

There are a few points to observe when adding dimensions to the document of a solid model.

1. Switch to model space by using the MSPACE command, not by setting the system variable TILEMODE.
2. Place the dimension entities on the layer with the suffix "-DIM."
3. Set the UCS to VIEW such that the XY plane of the new UCS is parallel to the current floating viewport.

To place the dimensions in model space, use the MSPACE command.

<View>   **<Floating Model Space>**

Command: **MSPACE**

You will add a linear dimension to the top view. Select the top view at A (Figure 5.12) to make this viewport as the current floating viewport.

The dimensioning layer for this floating viewport is TOP-DIM, set it as the current layer.

<Data>   **<Layers...>**

Command: **DDLMODES**
Current layer: **TOP-DIM**

In order for the dimensions to be placed on a plane parallel to the display view, align the UCS to the viewport.

[UCS]   **[View UCS]**

Command: **UCS**
Origin/ZAxis/3point/OBject/View/X/Y/Z/Prev/Restore/Save/Del/?/<World>: **V**

After you have switched to the floating model space, selected a floating viewport, made current a designated layer, and set the XY plane to align with the display view, you can create a linear dimension. See Figure 5.13.

[Dimensioning]   **[Linear Dimension]**

Command: **DIMLINEAR**
First extension line origin or RETURN to select: **END** of **[Select B (Figure 5.12).]**
Second extension line origin: **END** of **[Select C (Figure 5.12).]**

Dimension line location (Text/Angle/Horizontal/Vertical/Rotated): **[Select D (Figure 5.12).]**

Dimension text: **[Enter]**

While you are working in model space, you can use the ZOOM command to zoom in and out whenever necessary. To reiterate, the SOLVIEW command creates a saved view for each floating viewport. The name of the saved view is the name of the viewport. After zooming and panning, you can restore the display to the set view by running the VIEW command with the RESTORE option. If you have to adjust the size of any floating viewport that is created by the SOLVIEW command, you must update the saved view with the SAVE option of the VIEW command before you can zoom or pan.

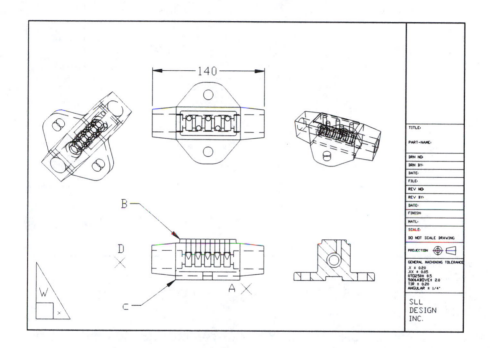

Figure 5.13 Linear dimension created in the top floating viewport

A dimension is created in the top viewport. To continue, you will make another linear dimension in the front view. Select the front view at A (Figure 5.13). Then, set the current layer to FRONT-DIM, align the UCS to the viewport, and create another linear dimension. See Figure 5.14.

    <Data>   **<Layers...>**

Command: **DDLMODES**
Current layer: **FRONT-DIM**

    [UCS]   **[View UCS]**

Command: **UCS**
Origin/ZAxis/3point/OBject/View/X/Y/Z/Prev/Restore/Save/Del/?/<World>: **V**

[Dimensioning]   [Linear Dimension]

Command: **DIMLINEAR**
First extension line origin or RETURN to select: **END** of **[Select B (Figure 5.13).]**
Second extension line origin: **END** of **[Select C (Figure 5.13).]**
Dimension line location (Text/Angle/Horizontal/Vertical/Rotated): **[Select D (Figure 5.13).]**

Dimension text: **[Enter]**

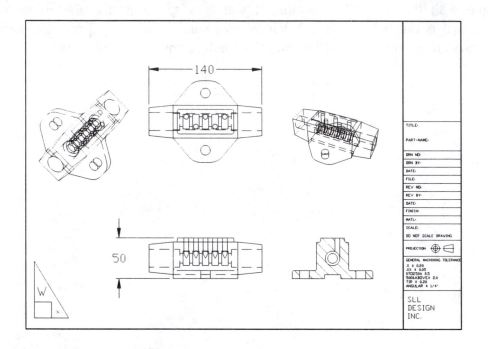

Figure 5.14 Dimension created in the front floating viewport

Now, two dimensions are created. You can continue to add other dimensions to the document.

After you have finished all the dimensions, do not forget to return to the paper space.

<View>   <Paper Space>

Command: **PSPACE**

The engineering document for the hydraulic valve is complete.

5.5 Summary

In this chapter, you learned the following AutoCAD commands related to 3D solid modeling.

SOLVIEW   SOLDRAW   SOLPROF

For a brief explanation of these commands, refer to the appendix of this book.

By now, you should be able to prepare a document from a solid model. Documentation includes setting up a title block, making floating viewports, and adding dimensions. In the next chapter, you will practice how to work in an open 3D space to create a structural framework.

5.6 Exercises

Following the steps outlined in this chapter, create the engineering documents for the drawing files that you created in Chapters 2, 3, and 4. Open the files one by one, set the TILEMODE system variable to 0. Then, make a layer and insert a proper title block drawing. Load the linetype HIDDEN to your drawing. After that, create a number of floating viewports by using the SOLVIEW command. Then, generate the visible lines, hidden lines, and the hatching lines by using the SOLDRAW command. Finally, add the dimensions accordingly. Figure 5.15 shows the completed engineering document prepared for the exercise in Chapter 4.

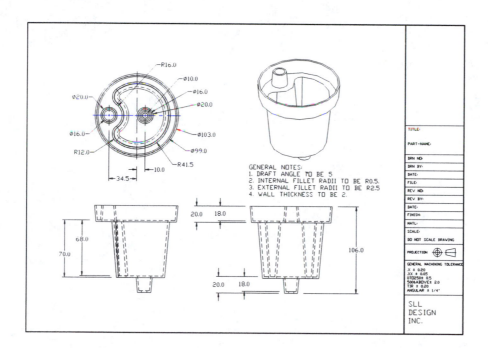

Figure 5.15 Engineering document for the casing of the water filtration unit

Chapter 6
Solid Modeling in 3D Space

The complex solid models that you created in Chapters 2 through 4 have one thing in common — the primitive solid features that make up the complex solid have close proximity.

In this chapter, you will build a 3D structural framework in an open space. The completed framework is shown in Figure 6.1. Structurally speaking, the construction of this framework is simple because it consists of only two types of solid features: cylinders and boxes. However, all the features must be positioned correctly in relation to each other in the 3D space. In making this model, you will practice how to manipulate the UCS and the objects in 3D space.

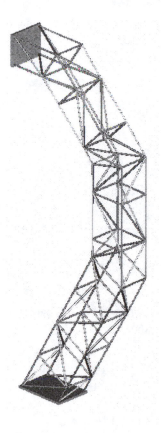

Figure 6.1 3D framework

Start a new drawing by using the NEW command.

<File>   <New...>

Command: **NEW**

Use ACADISO.DWG as the prototype drawing.

6.1   General Outline of the Framework

Although the framework looks very simple, careful planning before starting work is needed. Refer to Figure 6.1. You can divide the framework into four sections. To build the model systematically, you will need four layers to hold the entities of each section.

Within each section, and among the four sections themselves, the cylindrical solids should be properly aligned with each other. To simplify calculations and locating objects, you can create some general construction lines on the drawing to assist with alignment and judgment. You will need another layer for holding these construction lines.

Use the DDLMODES command to create five layers as follows:

<Data>   <Layers...>

Command: **DDLMODES**

Layer	Color
FRAME1	**yellow**
FRAME2	**green**
FRAME3	**magenta**
FRAME4	**red**
WIRE	**cyan**

Current layer: **WIRE**

The construction lines should lie on the ZX plane of the WCS (World Coordinate System), but the default working plane is the XY plane of the WCS. Therefore, run the UCS command to set the UCS to rotate about the X axis for 90°.

[UCS]   [X Axis Rotate UCS]

Command: **UCS**
Origin/ZAxis/3point/OBject/View/X/Y/Z/Prev/Restore/Save/Del/?/<World>: **X**
Rotation angle about X axis: **90**

For the sake of convenience in creating construction lines, set the display to the plan view of the new UCS. Run the PLAN command.

<View>   <3D Viewpoint Presets>   <Plan View>   <Current>

Command: **PLAN**
<Current UCS>/Ucs/World: **[Enter]**

Place the UCS icon at the origin position. Run the UCSICON command.

<Options>   **<UCS>**   **<Icon Origin>**

Command: **UCSICON**
ON/OFF/All/Noorigin/ORigin <ON>: **OR**

With the current layer set to WIRE, create a series of line segments to describe the outline of the framework. See Figure 6.2.

[Draw]   **[Line]**

Command: **LINE**
From point: **0,0**
To point: **@2000<60**
To point: **@1000<90**
To point: **@1000<135**
To point: **@400<180**
To point: **[Enter]**

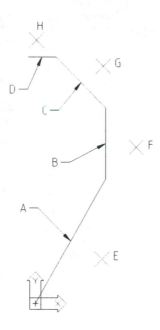

Figure 6.2 Series of line segments created

Create four offset lines using the OFFSET command. Then, run the XLINE command to create a horizontal line of infinite length. See Figure 6.3.

[Copy]   **[Offset]**

Command: **OFFSET**
Offset distance or Through <Through>: **450**
Select object to offset: **[Select A (Figure 6.2).]**
Side to offset? **[Select E (Figure 6.2).]**
Select object to offset: **[Enter]**

[Copy]   **[Offset]**

Command: **OFFSET**
Offset distance or Through <450.0000>: **400**
Select object to offset: **[Select B (Figure 6.2).]**
Side to offset? **[Select F (Figure 6.2).]**
Select object to offset: **[Enter]**

[Copy]   **[Offset]**

Command: **OFFSET**
Offset distance or Through <400.0000>: **350**
Select object to offset: **[Select C (Figure 6.2).]**
Side to offset? **[Select G (Figure 6.2).]**
Select object to offset: **[Enter]**

[Copy]   **[Offset]**

Command: **OFFSET**
Offset distance or Through <350.0000>: **300**
Select object to offset: **[Select D (Figure 6.2).]**
Side to offset? **[Select H (Figure 6.2).]**
Select object to offset: **[Enter]**

[Line]   **[Construction Line]**

Command: **XLINE**
Hor/Ver/Ang/Bisect/Offset/<From point>: **H**
Through point: **0,0**
Through point: **[Enter]**

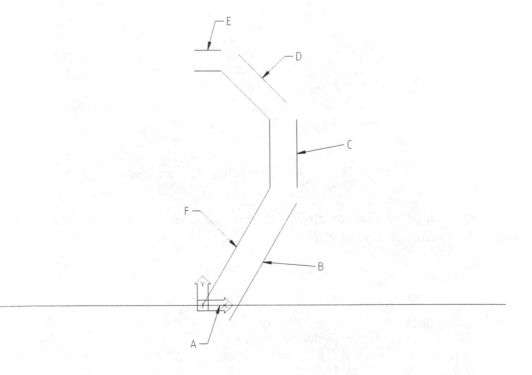

Figure 6.3 Offset lines and a xline created

Use the FILLET command to set the fillet radius to zero, and then fillet the corners of the line segments such that the outcome resembles Figure 6.4.

[Feature]   **[Fillet]**

Command: **FILLET**
(TRIM mode) Current fillet radius = 0.0000
Polyline/Radius/Trim/<Select first object>: **R**
Enter fillet radius <0.0000>: **0**

[Feature]   **[Fillet]**

Command: **FILLET**
(TRIM mode) Current fillet radius = 0.0000
Polyline/Radius/Trim/<Select first object>: **[Select A (Figure 6.3).]**
Select second object: **[Select B (Figure 6.3).]**

[Feature]   **[Fillet]**

Command: **FILLET**
(TRIM mode) Current fillet radius = 0.0000
Polyline/Radius/Trim/<Select first object>: **[Select B (Figure 6.3).]**
Select second object: **[Select C (Figure 6.3).]**

[Feature]   **[Fillet]**

Command: **FILLET**
(TRIM mode) Current fillet radius = 0.0000
Polyline/Radius/Trim/<Select first object>: **[Select C (Figure 6.3).]**
Select second object: **[Select D (Figure 6.3).]**

[Feature]   **[Fillet]**

Command: **FILLET**
(TRIM mode) Current fillet radius = 0.0000
Polyline/Radius/Trim/<Select first object>: **[Select D (Figure 6.3).]**
Select second object: **[Select E (Figure 6.3).]**

[Feature]   **[Fillet]**

Command: **FILLET**
(TRIM mode) Current fillet radius = 0.0000
Polyline/Radius/Trim/<Select first object>: **[Select A (Figure 6.3).]**
Select second object: **[Select F (Figure 6.3).]**

Figure 6.4 Lines filleted to zero radii

The construction lines required are completed. Run the UCS command to reset the UCS back to WORLD.

[UCS]   **[World UCS]**

Command: **UCS**
Origin/ZAxis/3point/OBject/View/X/Y/Z/Prev/Restore/Save/Del/?/<World>: **W**

6.2   First Section of the Framework

To obtain an isometric view, set the viewing position to rotate 315° in the XY plane, and 25° from the XY plane.

<View>   **<3D Viewpoint>**   **<Vector>**

Command: **VPOINT**
Rotate/<View point>: **R**
Enter angle in XY plane from X axis: **315**
Enter angle from XY plane: **25**

You will start working on the first section of the framework. Set the current layer to FRAME1. Then, lock the layer WIRE, so that you will not accidentally move the entities on this layer.

<Data>   **<Layers...>**

Command: **DDLMODES**

Layer	
WIRE	**Lock**

Current layer: **FRAME1**

Start the modeling by using the CYLINDER command and the ARRAY command to create four solid cylinders. See Figure 6.5.

[Solids]   **[Cylinder]**   **[Center]**

Command: **CYLINDER**
Elliptical/<center point>: **0,0,0**
Diameter/<Radius>: **10**
Center of other end/<Height>: **2100**

[Copy]   **[Rectangular Array]**

Command: **ARRAY**
Select objects: **LAST**
Select objects: **[Enter]**
Rectangular or Polar array (R/P): **R**
Number of rows (---) : **2**
Number of columns (||||) : **2**
Unit cell or distance between rows (---): **350**
Distance between columns (||||): **450**

Figure 6.5 Four cylinders created

The four cylinders created are the main supporting members of the framework. You need eight orthogonal cross members to link up the main members.

Use the CYLINDER command to create another solid cylinder. Then, run the COPY command to make a copy of this cross member.

[Solids]   **[Cylinder]**   **[Center]**

Command: **CYLINDER**
Elliptical/<center point>: **0,0,0**
Diameter/<Radius>: **10**
Center of other end/<Height>: **C**
Center of other end: **@450<0**

[Modify]   **[Copy Object]**

Command: **COPY**
Select objects: **LAST**
Select objects: **[Enter]**
<Base point or displacement>/Multiple: **350<90**
Second point of displacement: **[Enter]**

Repeat the CYLINDER command to produce one more cylinder. Also, make a copy of this cylinder by using the COPY command.

Now, you have four cross members. See Figure 6.6.

[Solids]   **[Cylinder]**   **[Center]**

Command: **CYLINDER**
Elliptical/<center point>: **0,0,0**
Diameter/<Radius>: **10**
Center of other end/<Height>: **C**
Center of other end: **@350<90**

[Modify]   **[Copy Object]**

Command: **COPY**
Select objects: **LAST**
Select objects: **[Enter]**
<Base point or displacement>/Multiple: **450<0**
Second point of displacement: **[Enter]**

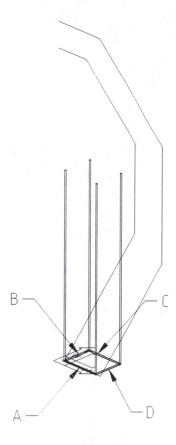

Figure 6.6 Four orthogonal cross members created

The four cross members should be 500 units away from the lower end of the vertical members. Run the MOVE command to translate them a distance of 500 units in the Z direction.

[Modify]   **[Move]**

Command: **MOVE**
Select objects: **[Select A, B, C, and D (Figure 6.6).]**
Select objects: **[Enter]**
Base point or displacement: **0,0,500**
Second point of displacement: **[Enter]**

Because there are two sets of orthogonal cross members, use the 3DARRAY command to create an array of two levels. See Figure 6.7.

[Copy]   **[3D Rectangular Array]**

```
Command: 3DARRAY
Select objects: P
Select objects: [Enter]
Rectangular or Polar array (R/P): R
Number of rows (---): 1
Number of columns (|||): 1
Number of levels (...): 3
Distance between levels (...): 500
```

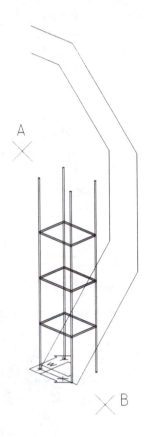

Figure 6.7 Orthogonal cross members moved and arrayed

The 16 cylinders you created form the basic shape of the first section of the framework. This section should be inclined at an angle. Use the ROTATE3D command to rotate these cylinders for 30° about the Y axis. See Figure 6.8.

[Modify]   **[3D Rotate]**

```
Command: ROTATE3D
Select objects: [Select A (Figure 6.7).]
Other corner: [Select B (Figure 6.7).]
Select objects: [Enter]
Axis by Object/Last/View/Xaxis/Yaxis/Zaxis/<2points>: Y
Point on Y axis: 0,0,0
<Rotation angle>/Reference: 30
```

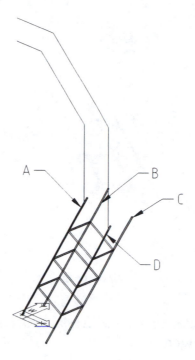

Figure 6.8 Framework rotated

As you can see in Figure 6.8, the lengths of the main members have been made longer than required. This was done intentionally because the two ends have to be cut to a slant angle, and it is time-consuming to determine the exact length of the cylinders. To cut the lower ends, use the SLICE command. See Figure 6.9.

```
[Solids]              [Slice]

Command: SLICE
Select objects: [Select A, B, C, and D (Figure 6.8).]
Select objects: [Enter]
Slicing plane by Object/Zaxis/View/XY/YZ/ZX/<3points>: XY
Point on XY plane: 0,0,25
Both sides/<Point on desired side of the plane>: CEN of [Select C (Figure 6.8).]
```

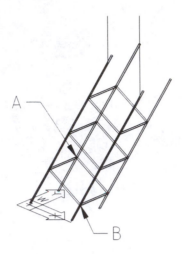

Figure 6.9 Lower ends of the main members sliced

To proceed, you will create the angular cross members for the framework. Use the CYLINDER command. See Figure 6.10.

[Solids]   **[Cylinder]**   **[Center]**

Command: **CYLINDER**
Elliptical/<center point>: **0,0,0**
Diameter/<Radius>: **10**
Center of other end/<Height>: **C**
Center of other end: **CEN** of **[Select A (Figure 6.9).]**

[Solids]   **[Cylinder]**   **[Center]**

Command: **CYLINDER**
Elliptical/<center point>: **0,0,0**
Diameter/<Radius>: **10**
Center of other end/<Height>: **C**
Center of other end: **CEN** of **[Select B (Figure 6.9).]**

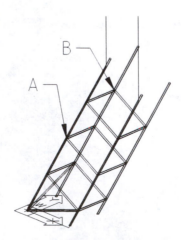

Figure 6.10 Two angular cross members created

Repeat the CYLINDER command to create one more cross member. See Figure 6.11.

[Solids]   **[Cylinder]**   **[Center]**

Command: **CYLINDER**
Elliptical/<center point>: **CEN** of **[Select A (Figure 6.10).]**
Diameter/<Radius>: **10**
Center of other end/<Height>: **C**
Center of other end: **CEN** of **[Select B (Figure 6.10).]**

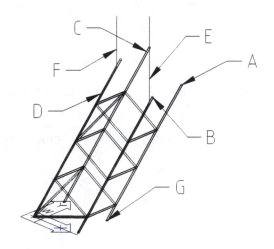

Figure 6.11 Another angular cross member created

Use the SLICE command to cut away the upper part of the four main members. See Figure 6.12.

[Solids]   **[Slice]**

Command: **SLICE**
Select objects: **[Select A, B, C, and D (Figure 6.11).]**
Select objects: **[Enter]**
Slicing plane by Object/Zaxis/View/XY/YZ/ZX/<3points>: **END** of **[Select E (Figure 6.11).]**
2nd point on plane: **END** of **[Select F (Figure 6.11).]**
3rd point on plane: **@1<90**
Both sides/<Point on desired side of the plane>: **CEN** of **[Select G (Figure 6.11).]**

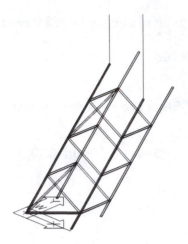

Figure 6.12 Upper ends of the main members sliced

Use the CYLINDER command to create the remaining cross members for the first section. See Figure 6.13.

[Solids]   **[Cylinder]**   **[Center]**

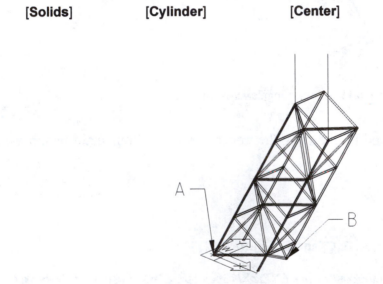

Figure 6.13   All cross members created

To complete the first section of the framework, run the BOX command to add a solid box. Then use the UNION command to unite all the members, including the box, to form a complex solid. See Figure 6.14.

[Solids]   **[Box]**   **[Corner]**

```
Command: BOX
Center/<Corner of box>: FROM
Base point: CEN of [Select A (Figure 6.13).]
<Offset>: @-25,-25
Cube/Length/<other corner>: FROM
Base point: CEN of [Select B (Figure 6.13).]
<Offset>: @25,25
Height: -25


[Modify]                        [Union]


Command: UNION
Select objects: [Select all the cylinders and the box.]
Select objects: [Enter]
```

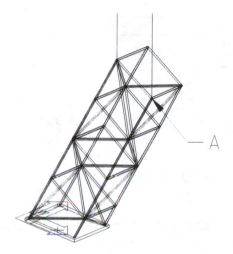

Figure 6.14 Solid box created and all members united

The first section of the framework is complete. Figure 6.15 shows a rendered view of the first section.

Figure 6.15 Rendered drawing of the first section

6.3   Second Section of the Framework

The second section of the framework is similar to the first section except that the main members are vertical. After creating the main members, you have to create the orthogonal cross members. Then, you have to cut the two ends of the main members at a slant angle to match the other sections. Finally, you will make the angular cross members and unite them.

Turn off the layer FRAME1, and set the current layer to layer FRAME2. You will construct the second section on this layer.

<Data>   **<Layers...>**

Command: **DDLMODES**

Layer	
FRAME1	**Off**

Current layer: **FRAME2**

Set the UCS to a new origin position by using the UCS command. Then, run the CYLINDER command to create one cylinder, and then run the ARRAY command to create three more cylinders. See Figure 6.16.

[UCS]   **[Origin UCS]**

Command: **UCS**
Origin/ZAxis/3point/OBject/View/X/Y/Z/Prev/Restore/Save/Del/?/<World>: **OR**
Origin point: **END** of **[Select A (Figure 6.14).]**

[Solids]   **[Cylinder]**   **[Center]**

Command: **CYLINDER**
Elliptical/<center point>: **0,0,-50**
Diameter/<Radius>: **10**
Center of other end/<Height>: **1400**

[Copy]   **[Rectangular Array]**

Command: **ARRAY**
Select objects: **LAST**
Select objects: **[Enter]**
Rectangular or Polar array (R/P): **R**
Number of rows (---): **2**
Number of columns (||||): **2**
Unit cell or distance between rows (---): **350**
Distance between columns (||||): **-400**

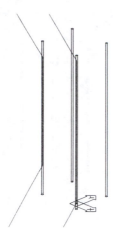

Figure 6.16 Main members of the second section created

Similar to the first section, create four orthogonal cross members. See Figure 6.17.

[Solids]   **[Cylinder]**   **[Center]**

Command: **CYLINDER**
Elliptical/<center point>: **0,0,0**
Diameter/<Radius>: **10**
Center of other end/<Height>: **C**
Center of other end: **@350<90**

[Modify]   **[Copy Object]**

Command: **COPY**
Select objects: **LAST**
Select objects: **[Enter]**
<Base point or displacement>/Multiple: **400<180**
Second point of displacement: **[Enter]**

[Solids]   **[Cylinder]**   **[Center]**

Command: **CYLINDER**
Elliptical/<center point>: 0,0,0
Diameter/<Radius>: 10
Center of other end/<Height>: **C**
Center of other end: **@400<180**

[Modify]   **[Copy Object]**

Command: **COPY**
Select objects: **LAST**
Select objects: **[Enter]**
<Base point or displacement>/Multiple: **350<90**
Second point of displacement: **[Enter]**

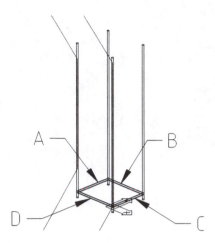

Figure 6.17 Four orthogonal cross members created

Use the MOVE command to translate the four orthogonal cross members for a distance of 400 units in the current Z direction. Then, use the 3DARRAY command to make a rectangular array of two levels of the cross members. The eight orthogonal cross members are complete. See Figure 6.18.

[Modify]   **[Move]**

Command: **MOVE**
Select objects: **[Select A, B, C, and D (Figure 6.17).]**
Select objects: **[Enter]**
Base point or displacement: **0,0,400**
Second point of displacement: **[Enter]**

[Modify]   **[3D Rectangular Array]**

Command: **3DARRAY**
Select objects: **P**
Select objects: **[Enter]**
Rectangular or Polar array (R/P): **R**
Number of rows (---): **1**
Number of columns (||||): **1**
Number of levels (...): **2**
Distance between levels (...): **500**

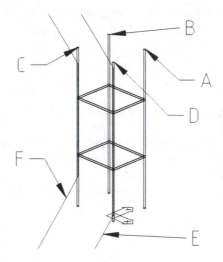

Figure 6.18 Orthogonal cross members moved and arrayed

Use the SLICE command to cut the lower edge of the four vertical members. See Figure 6.19.

[Solids]   **[Slice]**

Command: **SLICE**
Select objects: **[Select A, B, C, and D (Figure 6.18).]**
Select objects: **[Enter]**
Slicing plane by Object/Zaxis/View/XY/YZ/ZX/<3points>: **END** of **[Select E (Figure 6.18).]**
2nd point on plane: **END** of **[Select F (Figure 6.18).]**
3rd point on plane: **@1<90**
Both sides/<Point on desired side of the plane>: **CEN** of **[Select B (Figure 6.18).]**

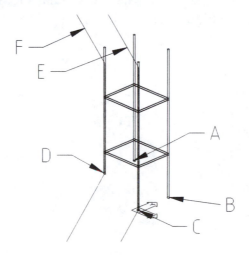

Figure 6.19 Lower ends of the main members sliced

Repeat the SLICE command to cut away the unwanted upper part of the vertical members. The vertical members are complete. See Figure 6.20.

[Solids]   **[Slice]**

Command: **SLICE**
Select objects: **[Select A, B, C, and D (Figure 6.19).]**
Select objects: **[Enter]**
Slicing plane by Object/Zaxis/View/XY/YZ/ZX/<3points>: **END** of **[Select E (Figure 6.19).]**
2nd point on plane: **END** of **[Select F (Figure 6.19).]**
3rd point on plane: **@1<90**
Both sides/<Point on desired side of the plane>: **CEN** of **[Select B (Figure 6.19).]**

Figure 6.20 Upper ends of the main members sliced

Run the CYLINDER command to create the angular cross members of the second section of the framework. See Figure 6.21.

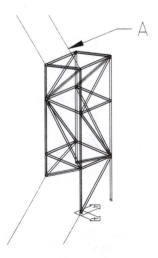

Figure 6.21 Angular cross members created

Run the UNION command to unite all the members of the second section of the framework. The second section of the framework is complete. Figure 6.22 shows a rendered view of the second section.

Figure 6.22 Rendered drawing of the second section

6.4   Third Section of the Framework

The third section is also similar to the first sections. Turn off the layer FRAME2 and set the current layer to layer FRAME3. You will place the third section on this layer.

<Data>   <Layers...>

Command: **DDLMODES**

Layer	
FRAME2	**Off**

Current layer: **FRAME3**

Set the origin of the UCS to a new position with the UCS command. Then, create four cylinders. See Figure 6.23.

[UCS]   [Origin UCS]

Command: **UCS**
Origin/ZAxis/3point/OBject/View/X/Y/Z/Prev/Restore/Save/Del/?/<World>: **OR**
Origin point: **END** of **[Select A (Figure 6.21).]**

[Solids]   [Cylinder]   [Center]

Command: **CYLINDER**
Elliptical/<center point>: **0,0,-50**
Diameter/<Radius>: **10**
Center of other end/<Height>: **1400**

[Copy]   [Rectangular Array]

Command: **ARRAY**
Select objects: **LAST**
Select objects: **[Enter]**
Rectangular or Polar array (R/P): **R**
Number of rows (---): **2**
Number of columns (||||): **2**
Unit cell or distance between rows (---): **350**
Distance between columns (||||): **-350**

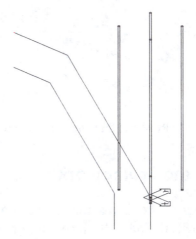

Figure 6.23 Four main members of the third section created

Create four orthogonal cross members for this section. See Figure 6.24.

[Solids]   **[Cylinder]**   **[Center]**

Command: **CYLINDER**
Elliptical/<center point>: **0,0,0**
Diameter/<Radius>: **10**
Center of other end/<Height>: **C**
Center of other end: **@350<90**

[Copy]   **[Polar Array]**

Command: **ARRAY**
Select objects: **L**
Select objects: **[Enter]**
Rectangular or Polar array (R/P): **P**
Center point of array: **-175,175**
Number of items: **4**
Angle to fill (+=ccw, -=cw): **360**
Rotate objects as they are copied? <Y> **[Enter]**

Figure 6.24 Four orthogonal members created

Run the MOVE command to move the four orthogonal cross members a distance of 400 units in the current Z direction. Then, create a rectangular array of them by using the 3DARRAY command. The distance between the levels in the array is 500 units. See Figure 6.25.

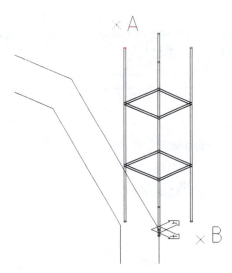

Figure 6.25 Orthogonal members moved and arrayed

Use the ROTATE3D command to rotate all the solid cylinders an angle of 45° about the current Y axis. See Figure 6.26.

[Modify]   **[3D Rotate]**

Command: **ROTATE3D**
Select objects: **[Select A (Figure 6.25).]**
Other corner: **[Select B (Figure 6.25).]**
Select objects: **[Enter]**
Axis by Object/Last/View/Xaxis/Yaxis/Zaxis/<2points>: **Y**
Point on Y axis: **0,0,0**
<Rotation angle>/Reference: **-45**

Figure 6.26 Framework rotated

Cut the ends of the four main members by using the SLICE command. Then, add angular cross members to the third section by using the CYLINDER command. After that, use the UNION command to unite all the members.

Set the UCS back to WORLD after completing the angular cross members. See Figure 6.27.

[UCS]   **[World UCS]**

Command: **UCS**
Origin/ZAxis/3point/OBject/View/X/Y/Z/Prev/Restore/Save/Del/?/<World>: **W**

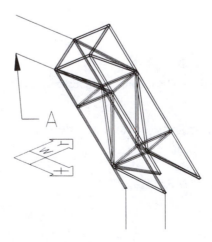

Figure 6.27 Completed third section of the framework

The third section of the framework is complete. Figure 6.28 shows the rendered view of this section.

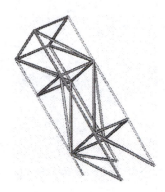

Figure 6.28 Rendered drawing of the third section

6.5 Completed Framework

Now we come to the final section of the framework. The method of construction is similar to that of the other sections.

Set the current layer to layer FRAME4 and turn off layer FRAME3. You will place the fourth section of the framework on layer FRAME4.

\<Data\>   \<Layers...\>

Command: **DDLMODES**

Layer	
FRAME3	**Off**

Current layer: **FRAME4**

Use the UCS command to set the UCS to align with the end point of the upper end of the construction line, and to set the Z axis of the UCS to point to the X direction. See Figure 6.29.

[UCS]   [Z Axis Vector UCS]

Command: **UCS**
Origin/ZAxis/3point/OBject/View/X/Y/Z/Prev/Restore/Save/Del/?/\<World\>: **ZA**
Origin point: **END** of **[Select A (Figure 6.27.]**
Point on positive portion of Z-axis: **@1,0**

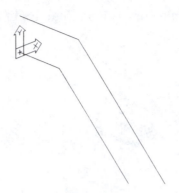

Figure 6.29 UCS set to a new position

Run the CYLINDER command and the ARRAY command to create four cylinders. See Figure 6.30.

[Solids]   **[Cylinder]**   **[Center]**

Command: **CYLINDER**
Elliptical/<center point>: **0,0,0**
Diameter/<Radius>: **10**
Center of other end/<Height>: **650**

[Copy]   **[Rectangular Array]**

Command: **ARRAY**
Select objects: **LAST**
Select objects: **[Enter]**
Rectangular or Polar array (R/P): **R**
Number of rows (---): **2**
Number of columns (||||): **2**
Unit cell or distance between rows (---): **300**
Distance between columns (||||): **350**

Figure 6.30 Main members of the fourth section created

Run the BOX command to produce a box. See Figure 6.31.

[Solids]   **[Box]**   **[Corner]**

Command: **BOX**
Center/<Corner of box>: **-25,-25**
Cube/Length/<other corner>: **FROM**
Base point: **CEN** of **[Select A (Figure 6.30]**
<Offset>: **@25,25**
Height: **-25**

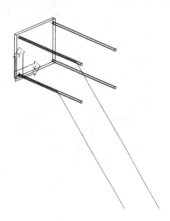

Figure 6.31 Box created

Cut the right ends of the cylinders and create the angular cross members. See Figure 6.32.

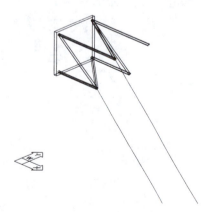

Figure 6.32 Ends cut away and the cross members created

Use the UNION command to unite the cylinders and the box to form the fourth section of the framework. Figure 6.33 shows the rendered view of this section.

Figure 6.33 Rendered drawing of the fourth section

You have completed the four sections of the 3D framework. Turn on all the layers and turn off layer WIRE. Finally, use the UNION command to unite the four sections. The 3D framework is complete. See Figure 6.34.

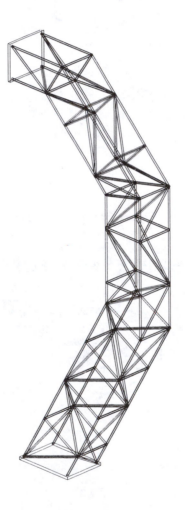

Figure 6.34 Completed 3D framework

6.6 Summary

In this chapter, you practiced the following commands related to working in 3D space.

3DARRAY   ROTATE3D   UCS

For a brief explanation of these commands, refer to the appendix of this book.

In this chapter, you have learned how to use simple construction lines to aid the manipulation of the UCS, and how to work in an open 3D space to build large spacious objects.

In the next chapter, you will learn how to apply the constructive solid geometry technique to architectural projects.

6.7   Exercise

To enhance your knowledge of creating 3D structural frameworks, you will create the solid model of an angle steel framework. The completed model is shown in Figure 6.35. This framework is made from angle steel bars of uniform cross section.

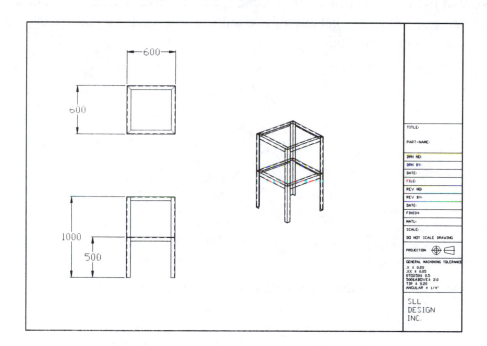

Figure 6.35 Angle steel framework

Start a new drawing to produce the framework. Make two additional layers, WIRE and SOLID, to hold the wireframe entities and solid entities, respectively. On the WIRE layer, make a 2D wireframe on the XY plane. See Figure 6.36.

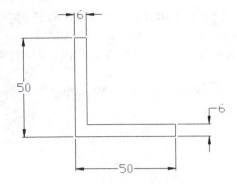

Figure 6.36 Cross section of the angle steel

Make two more copies of the cross section. Set the layer SOLID as the current layer. Next, extrude one of them a distance of 600 units. Then, rotate the extruded solid for 90° about the Y axis. See Figure 6.37.

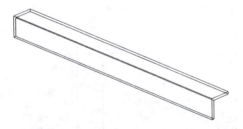

Figure 6.37 Two extruded solids created

Slice two ends of the angle steel at an angle of 45°. See Figure 6.38.

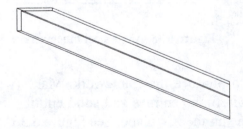

Figure 6.38 Both ends of the angle steel sliced

Extrude the second wireframe for a distance of 100 units. Then, copy the extruded solid. Next, translate the extruded solids to come in contact with the sliced solid. See Figure 6.39.

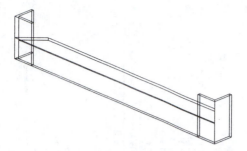

Figure 6.39 Two extruded solids translated

Subtract the two translated solids from the sliced solid. See Figure 6.40.

Figure 6.40 Sliced solid subtracted

Polar array the solid. See Figure 6.41.

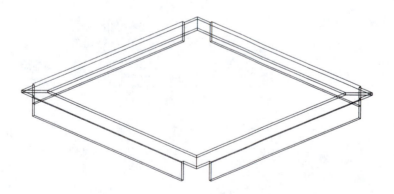

Figure 6.41 Solid arrayed

Make a 3D polar array of the four solids. Then add four extruded angle solids of 1200 length. See Figure 6.42.

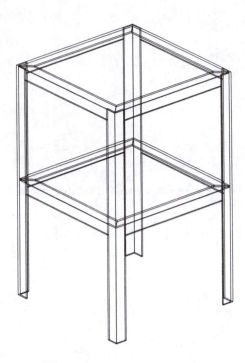

Figure 6.42 Completed angle steel framework

The angle steel framework is complete. Do not unite the solids. Otherwise, you will not be able to export each member of the framework individually to a drawing file to use as detail drawings for making each steel member.

Chapter 7
Architectural Project

Recall that a solid model in a computer is integrated mathematical data that contains information about the edges, surfaces, and volume of the object that the model describes. Basically, the set of constructive solid geometry commands of AutoCAD is an engineering design tool for mechanical and manufacturing engineering. Given such a useful and powerful tool, we should not limit its use to only the mentioned disciplines. In this chapter, you will use the constructive solid geometry modeling tool to create a 3D model of a house. The rendered drawing of the completed model is shown in Figure 7.1.

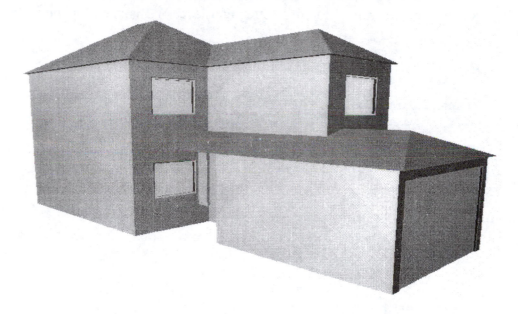

Figure 7.1 Rendered drawing of the completed building

The aims of this project are to explore and appreciate the use of constructive solid modeling techniques on architectural projects, to widen the scope of engineering application, and to broaden your perception of model creation. You might think of other applications, too.

In order to make this project easy to follow, the model has been simplified and many minute details have been omitted. This is not intended to imply that architectural projects should be simplified in such a way. After completing the model, you can, by all means, increase the detail of the model by adding more features to it.

Start a new drawing by using the NEW command.

<File>   <New...>

Command: **NEW**

Use ACADISO.DWG as the prototype drawing.

7.1   Analysis

Although the model of the house has been simplified and many details have been omitted, you must still build the internal floors, internal partitions, and the staircase, in addition to the exterior of the house.

You can divide the model into four major parts — the walls and floor of the first level, the staircase, the walls and floor of the second level, and the roofs.

Run the DDLMODES command to create six layers: F1, F2, STAIR, ROOF, W1, and W2. Layer F1 is used for the floor of the first level; layer F2 is used for the floor of the second level; layer STAIR is used for the staircase; layer ROOF is used for the roofs; layer W1 is used for the wall of the first level; and layer W2 is used for the wall of the second level.

<Data>   <Layers...>

Command: **DDLMODES**

Layer	Color
F1	yellow
F2	magenta
ROOF	blue
STAIR	red
W1	cyan
W2	green

Current layer: **W1**

7.2   Wireframes for the First Level

You will start with the first level of the house. On this level, there are two major object types: the walls and the floor. Figure 7.2 shows a dimensioned layout of the walls and floor plan of this level. You will create 2D wireframes for extrusion to 3D models. You will create wireframes first for the walls, and then for the floor. Check that the current layer is W1.

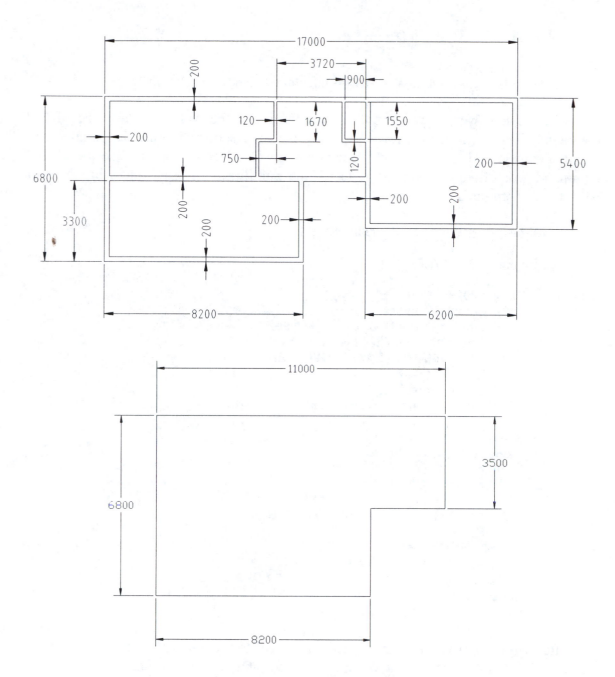

Figure 7.2 Dimensioned layout of the walls and the floor of the first level

Use the LIMITS command to set the limits to 20000 units times 10000 units. Then, run the ZOOM command to zoom to the extent of the drawing.

<Data>   **<Drawing Limits>**

Command: **LIMITS**
Reset Model space limits:
ON/OFF/<Lower left corner>: **0,0**
Upper right corner: **20000,10000**

[Zoom]   [Zoom Extents]

Command: **ZOOM**
All/Center/Dynamic/Extents/Left/Previous/Vmax/Window/<Scale(X/XP)>: **E**

A very quick way to create the 2D wireframes of the walls is to use the MLINE command. This command allows you to produce a series of multiple parallel lines simultaneously, and is very suitable for creating walls. Before you can use the MLINE command, you should set up the styles of the multilines, and save them by using the MLSTYLE command.

<Data>   <Multiline Style...>

Command: **MLSTYLE**

```
[Name: 200WALL
    [Element Properties...
    Elements:  Offset  Color        Ltype
               100     BYLAYER BYLAYER
               -100    BYLAYER BYLAYER
    OK                                          ]
    [Multiline Properties...
    Caps:      Line    Start   End
    OK                                          ]
Save...
Name: 120WALL
    [Element Properties...
    Elements:  Offset  Color        Ltype
               60 BYLAYER  BYLAYER
               -60     BYLAYER BYLAYER
    OK                                          ]
    [Multiline Properties...
    Caps:  Line        Start   End
    OK                                          ]
Save...
OK                                              ]
```

After setting up two multiline styles, run the MLINE command to create a multiline. Select the 200WALL style that you have created, and set the justification to TOP. See Figure 7.3.

[Polyline]   [Multiline]

Command: **MLINE**
Justification = Top, Scale = 1.00, Style = STANDARD
Justification/Scale/STyle/<From point>: **ST**
Mstyle name (or ?): **200WALL**
Justification/Scale/STyle/<From point>: **J**
Top/Zero/Bottom: **T**
Justification/Scale/STyle/<From point>: **0,0**
<To point>: **@6800<90**
Undo/<To point>: **@17000<0**
Close/Undo/<To point>: **@5400<270**

```
Close/Undo/<To point>: @6200<180
Close/Undo/<To point>: @5400<90
Close/Undo/<To point>: [Enter]
```

A multiline, as its name implies, has a number of parallel lines. Before you create a multiline, you need to decide which element of the multiline will coincide with the cursor selection point. The setting is governed by the method of justification. If you choose top justification, the uppermost multiline element will align with the selected position of the cursor. If you choose zero justification, the zero offset multiline element will align with the cursor. Finally, if you choose bottom justification, the lowermost multiline element will align with the cursor.

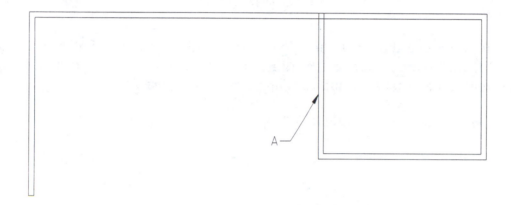

Figure 7.3 Set of multilines created

Repeat the MLINE command to create two more multilines. In making these lines, use bottom justification. See Figure 7.4.

```
[Polyline]                   [Multiline]

Command: MLINE
Justification = Top, Scale = 1.00, Style = 200WALL
Justification/Scale/STyle/<From point>: J
Top/Zero/Bottom: B
Justification/Scale/STyle/<From point>: 0,3300
Close/Undo/<To point>: PERP of [Select A (Figure 7.3).]
Close/Undo/<To point>: [Enter]

[Polyline]                   [Multiline]

Command: MLINE
Justification = Bottom, Scale = 1.00, Style = 200WALL
Justification/Scale/STyle/<From point>: 0,0
<To point>: @8200<0
Undo/<To point>: @3300<90
Close/Undo/<To point>: [Enter]
```

Compare the bottom justification with the top justification, and note the difference in multiline element location.

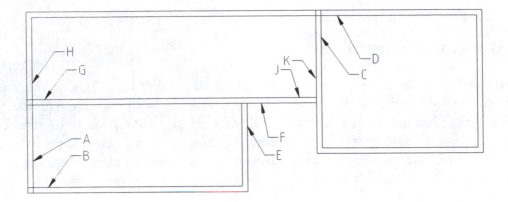

Figure 7.4 Two more multilines created

You have created three multilines. Now, you need to edit the joints between them. Run the MLEDIT command to edit a corner joint and four open tee joints. In making the open tee joints, the selection sequence is important. See Figure 7.5.

[Special Edit]   **[Edit Multiline]**

Command: **MLEDIT**

   [Corner Joint   ]

Select first mline: **[Select A (Figure 7.4).]**
Select second mline: **[Select B (Figure 7.4).]**
Select first mline(or Undo): **[Enter]**

[Special Edit]   **[Edit Multiline]**

Command: **MLEDIT**

   [Open Tee   ]

Select first mline: **[Select C (Figure 7.4).]**
Select second mline: **[Select D (Figure 7.4).]**
Select first mline(or Undo): **[Select E (Figure 7.4).]**
Select second mline: **[Select F (Figure 7.4).]**
Select first mline: **[Select G (Figure 7.4).]**
Select second mline: **[Select H (Figure 7.4).]**
Select first mline(or Undo): **[Select J (Figure 7.4).]**
Select second mline: **[Select K (Figure 7.4).]**
Select first mline(or Undo): **[Enter]**

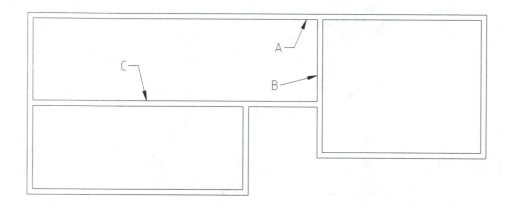

Figure 7.5 Five joints of the multilines edited

You have created the main walls of the first level. Run the MLINE command again to create the thinner walls. This time, use the 120WALL style. See Figure 7.6.

[Polyline]   **[Multiline]**

Command: **MLINE**
Justification = Bottom, Scale = 1.00, Style = 200WALL
Justification/Scale/STyle/<From point>: **ST**
Mstyle name (or ?): **120WALL**
Justification/Scale/STyle/<From point>: **J**
Top/Zero/Bottom: **T**
Justification/Scale/STyle/<From point>: **FROM**
Base point: **END** of **[Select A (Figure 7.5).]**
<Offset>: **@900<180**
<To point>: **@1550<270**
Undo/<To point>: **PERP** to **[Select B (Figure 7.5).]**
Close/Undo/<To point>: **[Enter]**

[Polyline]   **[Multiline]**

Command: **MLINE**
Justification = Top, Scale = 1.00, Style = 120WALL
Justification/Scale/STyle/<From point>: **FROM**
Base point: **END** of **[Select A (Figure 7.5).]**
<Offset>: **@3720<180**
<To point>: **@1670<270**
Undo/<To point>: **@750<180**
Close/Undo/<To point>: **PERP** to **[Select C (Figure 7.5).]**
Close/Undo/<To point>: **[Enter]**

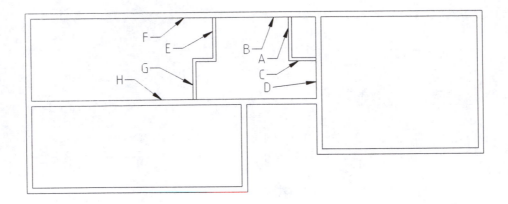

Figure 7.6 Two more multilines created

Again, you need to edit the joints between the multilines. Run the MLEDIT command. The wireframes of the first level are complete. See Figure 7.7.

[Special Edit]   **[Edit Multiline]**

Command: **MLEDIT**

	First mline	Second mline
Open Tee	A	B
Open Tee	C	D
Open Tee	E	F
Open Tee	G	H

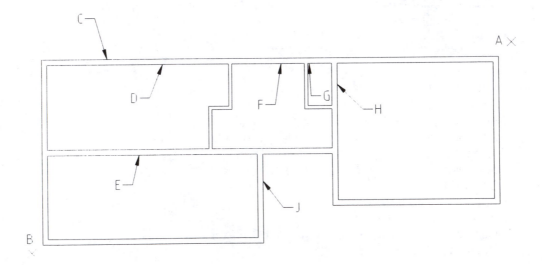

Figure 7.7 Multilines edited

If you want a 2D floor plan, then the first level is complete. Because you want to extrude the wireframes to a 3D model, you have to explode the multilines to separate line segments using the EXPLODE command.

[Modify]   **[Explode]**

Command: **EXPLODE**
Select objects: **[Select A (Figure 7.7).]**
Other corner: **[Select B (Figure 7.7).]**
Select objects: **[Enter]**

The system variable DELOBJ determines whether or not the original objects are deleted after they are used in an operation. Set its value to 1, so that the originals are deleted.

Command: **DELOBJ**
New value for DELOBJ: **1**

Form regions from the line segments by using the REGION command. Then, use the SUBTRACT command to create a complex region for extrusion to become a 3D solid model.

[Polygon]   **[Region]**

Command: **REGION**
Select objects: **[Select A (Figure 7.7).]**
Other corner: **[Select B (Figure 7.7).]**
Select objects: **[Enter]**
6 loops extracted.
6 Regions created.

[Modify]   **[Subtract]**

Command: **SUBTRACT**
Select solids and regions to subtract from...
Select objects: **[Select C (Figure 7.7).]**
Select objects: **[Enter]**
Select solids and regions to subtract...
Select objects: **[Select D (Figure 7.7).]**
Select objects: **[Select E (Figure 7.7).]**
Select objects: **[Select F (Figure 7.7).]**
Select objects: **[Select G (Figure 7.7).]**
Select objects: **[Select H (Figure 7.7).]**
Select objects: **[Enter]**

After exploding, region making, and Boolean operation, you should see no visual changes on the screen. The drawing on the screen should be the same as before.

The wireframe for the walls of the first level are complete. To continue with the first level, set the current layer to F1 by using the LAYER command. You will create a 2D region for the floor. This region should also lie on the XY plane of the WCS.

<Data>   <Layers...>

Command: **DDLMODES**

Layer	
F1	On
F2	On
ROOF	On
STAIR	On
W1	On
W2	On

Current layer: **F1**

Run the RECTANG command to create two rectangles. See Figure 7.8.

[Polygon]   **[Rectangle]**

Command: **RECTANG**
First corner: **0,0**
Other corner: **8200,6800**

[Polygon]   **[Rectangle]**

Command: **RECTANG**
First corner: **END** of **[Select J (Figure 7.7).]**
Other corner: **.X** of **END** of **[Select H (Figure 7.7).]**
(need YZ): **END** of **[Select C (Figure 7.7).]**

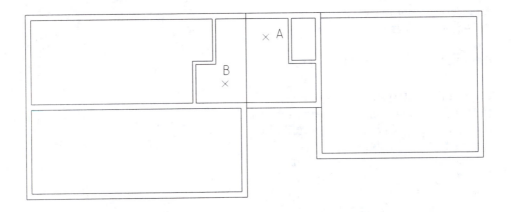

Figure 7.8 Two rectangles created

Run the REGION command to convert the rectangles into regions. Then, run the UNION command to unite them to form a complex region. See Figure 7.9.

[Polygon]   **[Region]**

Command: **REGION**
Select objects: **[Select A (Figure 7.8).]**
Other corner: **[Select B (Figure 7.8).]**
Select objects: **[Enter]**
2 loops extracted.
2 Regions created.

[**Modify**]   [**Union**]

Command: **UNION**
Select objects: **[Select A (Figure 7.8).]**
Other corner: **[Select B (Figure 7.8).]**
Select objects: **[Enter]**

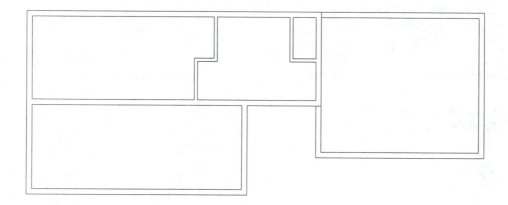

Figure 7.9 Two rectangles united

The 2D wireframes for the walls and floor of the first floor are complete. You will run the EXTRUDE command on them later.

7.3 Wireframes for the Staircase

The staircase runs in three directions. You can divide it into three sections and create three sets of wireframes. With the wireframes, you will form three extruded solids and unite them to form a model of the staircase. Set the current layer to STAIR, which will hold the wireframes for the staircase.

<Data>   <Layers...>

Command: **DDLMODES**

Layer	
F1	On
F2	On
ROOF	On
STAIR	On
W1	On
W2	On

Current layer: **STAIR**

Unlike the wireframes for the walls and floor that are 2D, the wireframes for the staircase are 3D. Therefore, you should set the viewing position to a 3D view. Run the VPOINT command to rotate the viewpoint 315° on the XY plane, and 25° from the XY plane. Place the UCS icon to display at the origin position by using the UCSICON command. See Figure 7.10.

<View>   **<3D Viewpoint>**   **<Vector>**

Command: **VPOINT**
Rotate/<View point>: **R**
Enter angle in XY plane from X axis: **315**
Enter angle from XY plane: **25**
Regenerating drawing.

<Options>   **<UCS>**   **<Icon Origin>**

Command: **UCSICON**
ON/OFF/All/Noorigin/ORigin: **OR**

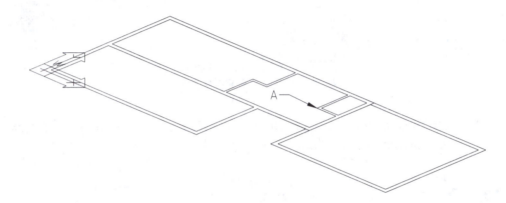

Figure 7.10 New viewing position

Set the UCS to a new position by using the UCS command. Place the origin at a point 180 units in the Z direction of point A (Figure 7.10), and set the orientation of the Z axis to point to the current X direction. See Figure 7.11.

[UCS]   **[Z Axis Vector UCS]**

Command: **UCS**
Origin/ZAxis/3point/OBject/View/X/Y/Z/Prev/Restore/Save/Del/?/<World>: **ZA**
Origin point: **FROM**
Base point: **END** of **[Select A (Figure 7.10).]**
<Offset>: **@0,0,180**
Point on positive portion of Z-axis: **@1,0**

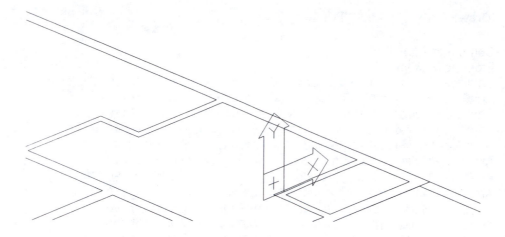

Figure 7.11 New UCS position

For your reference, the dimensions of the three sections of the staircase are shown in Figure 7.12.

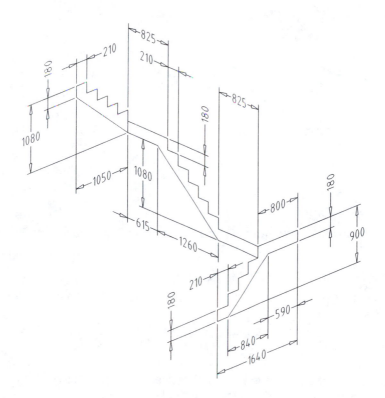

Figure 7.12 Staircase in three sections

Using the dimensions shown in Figure 7.12, run the PLINE command to create a polyline for the first section of the staircase. See Figure 7.13.

[Draw]   **[Polyline]**

Command: **PLINE**
From point: **30,0**
Arc/Close/Halfwidth/Length/Undo/Width/<Endpoint of line>: **@180<90**
Arc/Close/Halfwidth/Length/Undo/Width/<Endpoint of line>: **@210<0**
Arc/Close/Halfwidth/Length/Undo/Width/<Endpoint of line>: **@180<90**
Arc/Close/Halfwidth/Length/Undo/Width/<Endpoint of line>: **@210<0**
Arc/Close/Halfwidth/Length/Undo/Width/<Endpoint of line>: **@180<90**
Arc/Close/Halfwidth/Length/Undo/Width/<Endpoint of line>: **@210<0**
Arc/Close/Halfwidth/Length/Undo/Width/<Endpoint of line>: **@180<90**
Arc/Close/Halfwidth/Length/Undo/Width/<Endpoint of line>: **@210<0**
Arc/Close/Halfwidth/Length/Undo/Width/<Endpoint of line>: **@180<90**
Arc/Close/Halfwidth/Length/Undo/Width/<Endpoint of line>: **@800<0**
Arc/Close/Halfwidth/Length/Undo/Width/<Endpoint of line>: **@180<270**
Arc/Close/Halfwidth/Length/Undo/Width/<Endpoint of line>: **@590<180**
Arc/Close/Halfwidth/Length/Undo/Width/<Endpoint of line>: **@-840,-720**
Arc/Close/Halfwidth/Length/Undo/Width/<Endpoint of line>: **C**

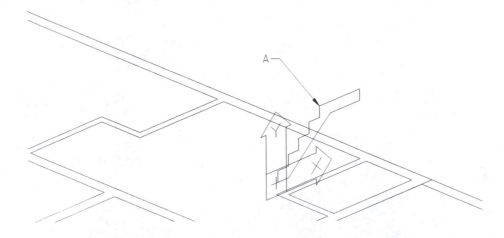

Figure 7.13 Wireframe for the first section of the staircase drawn

Set the UCS to a new position by using the UCS command. Then, run the PLINE command to create the wireframe for the second section of the staircase. See Figure 7.14.

[UCS]   **[Z Axis Vector UCS]**

Command: **UCS**
Origin/ZAxis/3point/OBject/View/X/Y/Z/Prev/Restore/Save/Del/?/<World>: **ZA**
Origin point: **END** of **[Select A (Figure 7.13).]**
Point on positive portion of Z-axis: **@-1,0**

[Draw]   **[Polyline]**

Command: **PLINE**
From point: **0,0**
Arc/Close/Halfwidth/Length/Undo/Width/<Endpoint of line>: **@825<180**

Arc/Close/Halfwidth/Length/Undo/Width/<Endpoint of line>: **@180<90**
Arc/Close/Halfwidth/Length/Undo/Width/<Endpoint of line>: **@210<180**
Arc/Close/Halfwidth/Length/Undo/Width/<Endpoint of line>: **@180<90**
Arc/Close/Halfwidth/Length/Undo/Width/<Endpoint of line>: **@210<180**
Arc/Close/Halfwidth/Length/Undo/Width/<Endpoint of line>: **@180<90**
Arc/Close/Halfwidth/Length/Undo/Width/<Endpoint of line>: **@210<180**
Arc/Close/Halfwidth/Length/Undo/Width/<Endpoint of line>: **@180<90**
Arc/Close/Halfwidth/Length/Undo/Width/<Endpoint of line>: **@210<180**
Arc/Close/Halfwidth/Length/Undo/Width/<Endpoint of line>: **@180<90**
Arc/Close/Halfwidth/Length/Undo/Width/<Endpoint of line>: **@210<180**
Arc/Close/Halfwidth/Length/Undo/Width/<Endpoint of line>: **@180<90**
Arc/Close/Halfwidth/Length/Undo/Width/<Endpoint of line>: **@825<180**
Arc/Close/Halfwidth/Length/Undo/Width/<Endpoint of line>: **@180<270**
Arc/Close/Halfwidth/Length/Undo/Width/<Endpoint of line>: **@615<0**
Arc/Close/Halfwidth/Length/Undo/Width/<Endpoint of line>: **@1260,-1080**
Arc/Close/Halfwidth/Length/Undo/Width/<Endpoint of line>: **@825<0**
Arc/Close/Halfwidth/Length/Undo/Width/<Endpoint of line>: **C**

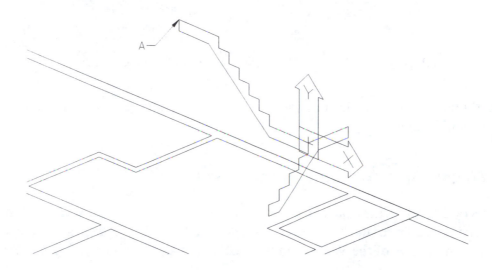

Figure 7.14 Wireframe for the second section of the staircase drawn

Set the UCS to a new position by using the UCS command again.

[UCS]   **[Z Axis Vector UCS]**

Command: **UCS**
Origin/ZAxis/3point/OBject/View/X/Y/Z/Prev/Restore/Save/Del/?/<World>: **ZA**
Origin point <0,0,0>: **END** of **[Select A (Figure 7.14).]**
Point on positive portion of Z-axis: **@1,0**

Using the dimensions shown in Figure 7.12, create the wireframe for the third section of the staircase. The outcome should resemble Figure 7.15.

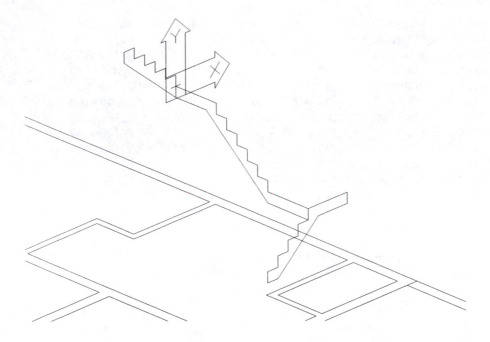

Figure 7.15 Wireframe for the third section of the staircase drawn

The 3D wireframes for the staircase of the house are complete. You will extrude them to become solids later.

7.4   Wireframes for the Second Level

On the second level of the house, there are also two object types: the walls and the floor. This level is similar to the first level, except that the XY plane lies on the positive direction of the Z axis of the WCS. Run the UCS command. See Figure 7.16.

[UCS]   **[Origin UCS]**

Command: **UCS**
Origin/ZAxis/3point/OBject/View/X/Y/Z/Prev/Restore/Save/Del/?/<World>: **OR**
Origin point: ***0,0,3060**

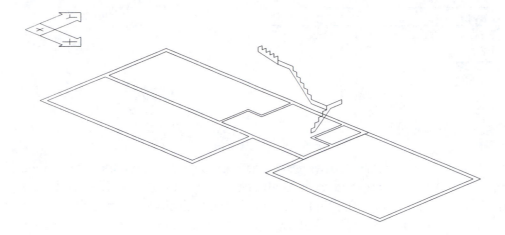

Figure 7.16 UCS position for the second level

You can follow the same approach for the first level to create the wireframes for the second level. Figure 7.17 shows a dimensioned layout of the second level.

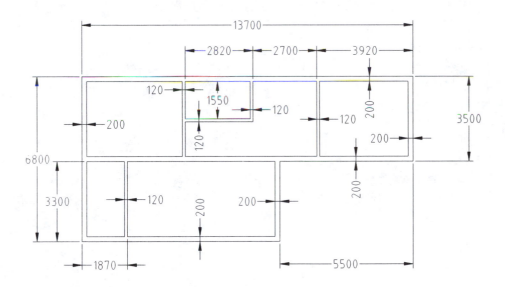

Figure 7.17 Dimensioned layout for the second level

Use the PLAN command to set the viewing direction to the plan view of the current UCS.

<View>   <3D Viewpoint Presets>   <Plan View>   <Current>

Set the current layer to W2. You will place the walls of the second floor on this layer.

<Data>   <Layers...>

Command: **DDLMODES**

Layer	
F1	On
F2	On
ROOF	On
STAIR	On
W1	On
W2	On

Current layer: **W2**

Refer to the dimensioned drawing to create the layout drawings for the second level. First, use the MLINE command to draw the necessary multilines. Then, edit the joints of the multilines by using the MLEDIT command. Next, explode the multilines into separate line segments by using the EXPLODE command. Finally, convert the line segments into a series of regions by using the REGION command, and run the SUBTRACT command on the regions to form a complex region. You will extrude the complex region later.

After making the wireframes for the walls, return the display to a 3D view by using the VPOINT command. Rotate the viewing angle 315° in the XY plane, and 25° from the XY plane. Check your drawing against Figure 7.18.

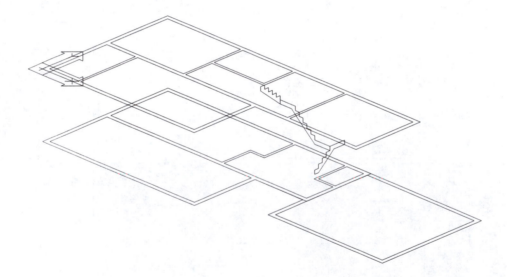

Figure 7.18 Multilines edited, exploded, and converted into regions and then to a complex region

The wireframes for the walls of the second level are complete. Set the current layer to F2 and turn off the layer W2.

<Data>   <Layers...>

Command: **DDLMODES**

Layer	
F1	**On**
F2	**On**
ROOF	**On**
STAIR	**On**
W1	**On**
W2	**Off**

Current layer: **F2**

To create the wireframes for the floor of the second level, use the RECTANG command to create three rectangles. See Figure 7.19.

[Polygon]   **[Rectangle]**

Command: **RECTANG**

Rectangle	First corner	Other corner
A	**0,0**	**8200,6800**
B	**8200,3300**	**13700,6800**
C	**7080,4960**	**9780,6800**

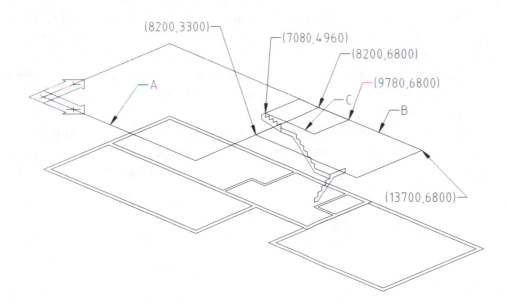

Figure 7.19 Three rectangles created

Use the REGION command to convert the rectangles to three regions. Then, run the UNION command to unite rectangles A and B (Figure 7.19) into a single region. Next, use the SUBTRACT command to subtract rectangle C (Figure 7.19) from the united region. The wireframe of the floor for the second level is complete. See Figure 7.20.

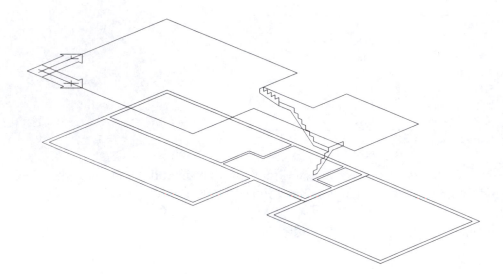

Figure 7.20 Wireframe for the floor of second level created

7.5 Wireframes for the Roofs

There are two roofs — one for the first level and another for the second level. The roof for the second level consists of two parts. In total, you will create seven wireframes in three groups. From the wireframes, you will create seven extruded solids. The roof of the first level is formed by the intersection of three extruded solids. The roof of the second level consists of two parts. Each part is the intersection of two extruded solids. The completed roof of the second level is the union of two solids of intersection.

Figure 7.21 shows the dimensions of the wireframes for the roofs. In the figure, the origin positions of the UCS for the three sets of wireframes are delineated in terms of WCS coordinates. For your reference, the lower-left corner of the floor of the first level is at (0,0,0) of the WCS.

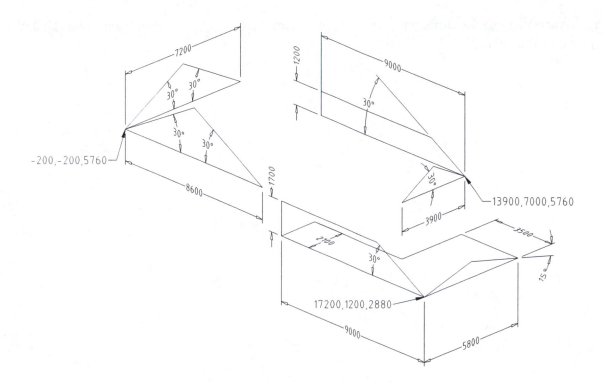

Figure 7.21 Dimensioned layout of the roofs

Set the current layer to ROOF, and turn off the layers STAIR, W1, W2, and F2. On your screen, you will have the entities on layer F1 remaining.

<Data>   <Layers...>

Command: **DDLMODES**

Layer	
F1	On
F2	Off
ROOF	On
STAIR	Off
W1	Off
W2	Off

Current layer: **ROOF**

Set the UCS to a new position by using the UCS command. The symbol (*) refers to the absolute WCS coordinates regardless of current UCS position.

[UCS]   [Z Axis Vector UCS]

Command: **UCS**
Origin/ZAxis/3point/OBject/View/X/Y/Z/Prev/Restore/Save/Del/?/<World>: **ZA**
Origin point: ***-200,-200,5760**
Point on positive portion of Z-axis: **@1,0**

According to the dimensions shown in Figure 7.21, create a 2D wireframe on the XY plane of the new UCS. See Figure 7.22.

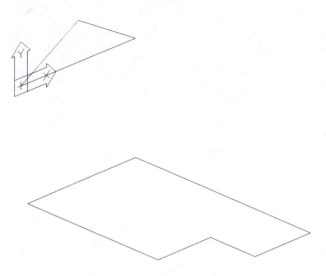

Figure 7.22 Wireframe for the roofs created

Set the UCS to rotate -90° about the current Y axis by using the UCS command.

[UCS]   **[Y Axis Rotate UCS]**

Command: **UCS**
Origin/ZAxis/3point/OBject/View/X/Y/Z/Prev/Restore/Save/Del/?/<World>: **Y**
Rotation angle about Y axis: **-90**

Draw another wireframe on this new UCS location. See Figure 7.23.

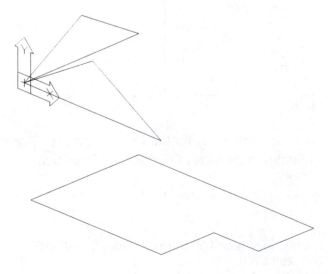

Figure 7.23 Second wireframe for the roofs created

Set the UCS location again for the second part of the roof for the second level.

[UCS]   [Z Axis Vector UCS]

Command: **UCS**
Origin/ZAxis/3point/OBject/View/X/Y/Z/Prev/Restore/Save/Del/?/<World>: **ZA**
Origin point: ***13900,7000,5760**
Point on positive portion of Z-axis: **@1,0**

Refer to the dimensioned drawing shown in Figure 7.21 to create the third wireframe. See Figure 7.24.

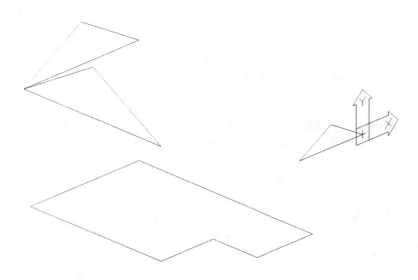

Figure 7.24 Third wireframe for the roofs created

Set the UCS again by using the UCS command. Rotate it -90° about the current Y axis. Then, create the fourth wireframe. See Figure 7.25.

[UCS]   [Y Axis Rotate UCS]

Command: **UCS**
Origin/ZAxis/3point/OBject/View/X/Y/Z/Prev/Restore/Save/Del/?/<World>: **Y**
Rotation angle about Y axis: **-90**

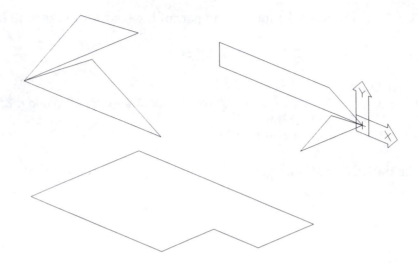

Figure 7.25 Fourth wireframe for the roofs created

Before proceeding to make the wireframes for the roof of the first level, turn off layer F1 so that only those entities related to the roof are left on the screen.

<Data>   <Layers...>

Command: **DDLMODES**

Layer	
F1	Off
F2	Off
ROOF	On
STAIR	Off
W1	Off
W2	Off

Current layer: **ROOF**

Switch to the roof for the first level. Just like for the other two sets of wireframes, set the UCS to a new position.

[UCS]   [Z Axis Vector UCS]

Command: **UCS**
Origin/ZAxis/3point/OBject/View/X/Y/Z/Prev/Restore/Save/Del/?/<World>: **ZA**
Origin point: ***17200,1200,2880**
Point on positive portion of Z-axis: **@1,0**

Refer to Figure 7.21 to create the fifth wireframe. See Figure 7.26.

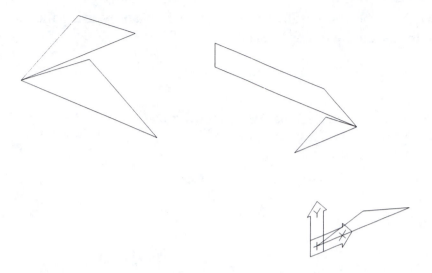

Figure 7.26 Fifth wireframe for the roofs created

Set the UCS to a new position again by using the UCS command.

[UCS]   **[Y Axis Rotate UCS]**

Command: **UCS**
Origin/ZAxis/3point/OBject/View/X/Y/Z/Prev/Restore/Save/Del/?/<World>: **Y**
Rotation angle about Y axis <0>: **-90**

Create the sixth wireframe. See Figure 7.27.

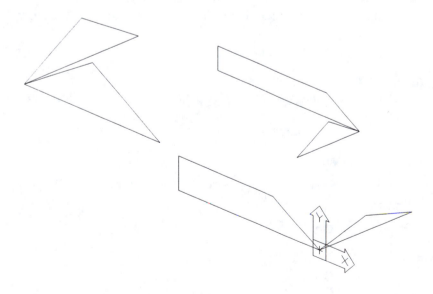

Figure 7.27 The sixth wireframe for the roofs created

Now switch to the last wireframe. Set the UCS again by using the UCS command. Then, create the seventh wireframe. See Figure 7.28.

[UCS]   **[X Axis Rotate UCS]**

Command: **UCS**
Origin/ZAxis/3point/OBject/View/X/Y/Z/Prev/Restore/Save/Del/?/<World>: **X**
Rotation angle about X axis <0>: **-90**

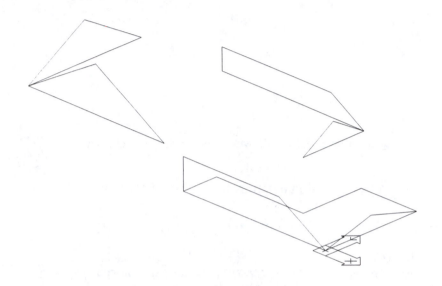

Figure 7.28 Wireframes for the roofs completed

All the wireframes for the model of the house are complete. Reset the UCS to WORLD by using the UCS command. Then, turn on all the layers. See Figure 7.29.

[UCS]   **[World UCS]**

Command: **UCS**
Origin/ZAxis/3point/OBject/View/X/Y/Z/Prev/Restore/Save/Del/?/<World>: **W**

<Data>   **<Layers...>**

Command: **DDLMODES**

Layer	
F1	On
F2	On
ROOF	On
STAIR	On
W1	On
W2	On

Current layer: **ROOF**

All the wireframes are complete.

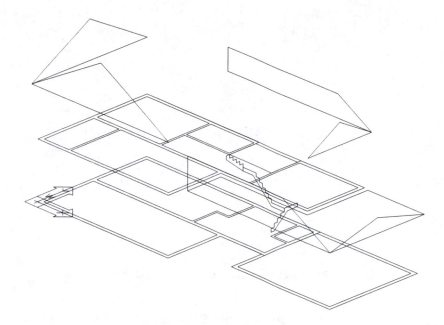

Figure 7.29 Wireframes required for the house

7.6   3D Model for the First Level and the Staircase

After finishing all the wireframes, you can create the model for the first level of the house. You will start by working on the staircase, the floor, and then the walls. Before advancing to the second level, you will cut openings for the windows and doors on the walls.

Set the current layer to layer STAIR, and turn off layers ROOF, W2, and F2. See Figure 7.30.

<Data>   **<Layers...>**

Command: **DDLMODES**

Layer	
F1	On
F2	Off
ROOF	Off
STAIR	On
W1	On
W2	Off

Current layer: **STAIR**

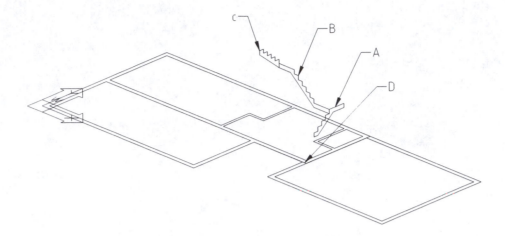

Figure 7.30 Wireframes for the first level and the staircase

The model for the staircase is divided into three sections, and you have created three wireframes for them. To make the staircase, you will extrude them to 3D solids. When you extrude a region or a polyline, the direction of extrusion is in the Z direction of the UCS where the region or polyline is created, not the current UCS. Therefore, refer to Figures 7.13, 7.14, and 7.15 to check the UCS of the object. For the first and the second sections, the extrusion should be in the negative direction. For the third section, the extrusion should be in the positive direction.

Run the EXTRUDE command to extrude the wireframes. Then use the UNION command to unite the three sections of staircase.

[Solids]   **[Extrude]**

Command: **EXTRUDE**
Select objects: **[Select A (Figure 7.30).]**
Select objects: **[Enter]**
Path/<Height of Extrusion>: **-825**
Extrusion taper angle <0>: **[Enter]**

[Solids]   **[Extrude]**

Command: **EXTRUDE**
Select objects: **[Select B (Figure 7.30).]**
Select objects: **[Enter]**
Path/<Height of Extrusion>: **-800**
Extrusion taper angle <0>: **[Enter]**

[Solids]   **[Extrude]**

Command: **EXTRUDE**
Select objects: **[Select C (Figure 7.30).]**
Select objects: **[Enter]**
Path/<Height of Extrusion>: **825**
Extrusion taper angle <0>: **[Enter]**

[Modify]   **[Union]**

Command: **UNION**
Select objects: **[Select the three extruded staircases.]**
Select objects: **[Enter]**

The staircase is complete. Set the current layer to layer F1. Then, run the EXTRUDE command to extrude the wireframe for the floor. See Figure 7.31.

<Data>   **<Layers...>**

Command: **DDLMODES**

Layer	
F1	On
F2	Off
ROOF	Off
STAIR	On
W1	On
W2	Off

Current layer: **F1**

[Solids]   **[Extrude]**

Command: **EXTRUDE**
Select objects: **[Select D (Figure 7.30).]**
Select objects: **[Enter]**
Path/<Height of Extrusion>: **180**
Extrusion taper angle <0>: **[Enter]**

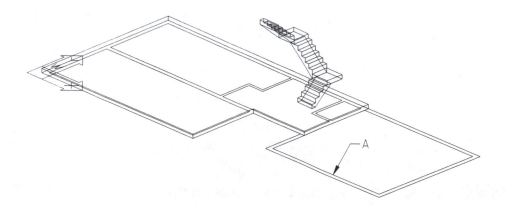

Figure 7.31 Floor and the staircase created

The floor for the first level is complete. Before making the walls, set the current layer to layer W1 and turn off the layer F1.

<Data>   **<Layers...>**

Command: **DDLMODES**

Layer	
F1	Off
F2	Off
ROOF	Off
STAIR	On
W1	On
W2	Off

Current layer: **W1**

Run the EXTRUDE command to create the walls of the first level. See Figure 7.32.

[Solids]   **[Extrude]**

Command: **EXTRUDE**
Select objects: **[Select A (Figure 7.31).]**
Select objects: **[Enter]**
Path/<Height of Extrusion>: **2880**
Extrusion taper angle <0>: **[Enter]**

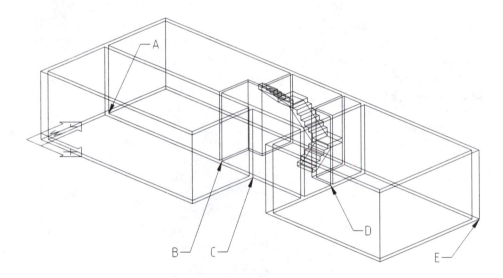

Figure 7.32 Walls created

The walls should have openings for the doors and windows. Figure 7.33 shows a dimensioned layout of these openings. The height of the doors and gate are 2000 units and 2600 units, respectively. The window opening is 1000 units from the top of the floor and is 1200 units high.

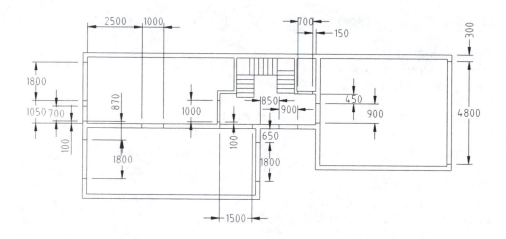

Figure 7.33 Dimensions for the windows and doors

Refer to the dimensions shown in Figure 7.33 and use the BOX command to prepare eight door openings. See Figure 7.34.

[Solids]   [Box]   [Corner]

Command: **BOX**
Center/<Corner of box>: **FROM**
Base point: **END** of **[Select A (Figure 7.32).]**
<Offset>: **@0,100,180**
Cube/Length/<other corner>: **@-200,700,2000**

[Solids]   [Box]   [Corner]

Command: **BOX**
Center/<Corner of box>: **FROM**
Base point: **END** of **[Select A (Figure 7.32).]**
<Offset>: **@2500,0,180**
Cube/Length/<other corner>: **@1000,-200,2000**

[Solids]   [Box]   [Corner]

Command: **BOX**
Center/<Corner of box>: **FROM**
Base point: **END** of **[Select B (Figure 7.32).]**
<Offset>: **@0,100,180**
Cube/Length/<other corner>: **@-120,1000,2000**

[Solids]   [Box]   [Corner]

Command: **BOX**
Center/<Corner of box>: **FROM**
Base point: **END** of **[Select B (Figure 7.32).]**
<Offset>: **@0,0,180**
Cube/Length/<other corner>: **@1500,-200,2000**

[Solids]   **[Box]**   **[Corner]**

Command: **BOX**
Center/<Corner of box>: **FROM**
Base point: **END** of **[Select C (Figure 7.32).]**
<Offset>: **@850,0,180**
Cube/Length/<other corner>: **@900,200,2000**

[Solids]   **[Box]**   **[Corner]**

Command: **BOX**
Center/<Corner of box>: **FROM**
Base point: **END** of **[Select D (Figure 7.32).]**
<Offset>: **@-150,0,180**
Cube/Length/<other corner>: **@-700,120,2000**

[Solids]   **[Box]**   **[Corner]**

Command: **BOX**
Center/<Corner of box>: **FROM**
Base point: **END** of **[Select D (Figure 7.32).]**
<Offset>: **@0,-450,180**
Cube/Length/<other corner>: **@200,-900,2000**

[Solids]   **[Box]**   **[Corner]**

Command: **BOX**
Center/<Corner of box>: **FROM**
Base point: **END** of **[Select E (Figure 7.32).]**
<Offset>: **@0,-300,0**
Cube/Length/<other corner>: **@-200,-4800,2600**

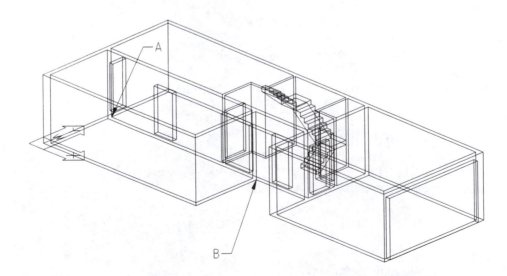

Figure 7.34 Openings for the doors created

Using the dimensions shown in Figure 7.33, run the BOX command to prepare three window openings. See Figure 7.35.

[Solids]   **[Box]**   **[Corner]**

Command: **BOX**
Center/<Corner of box>: **FROM**
Base point: **END** of **[Select A (Figure 7.34).]**
<Offset>: **@0,1050,1180**
Cube/Length/<other corner>: **@-200,1800,1200**

[Solids]   **[Box]**   **[Corner]**

Command: **BOX**
Center/<Corner of box>: **FROM**
Base point: **END** of **[Select A (Figure 7.34).]**
<Offset>: **@0,-770,1180**
Cube/Length/<other corner>: **@-200,-1800,1200**

[Solids]   **[Box]**   **[Corner]**

Command: **BOX**
Center/<Corner of box>: **FROM**
Base point: **END** of **[Select B (Figure 7.34).]**
<Offset>: **@0,-650,1180**
Cube/Length/<other corner>: **@-200,-1800,1200**

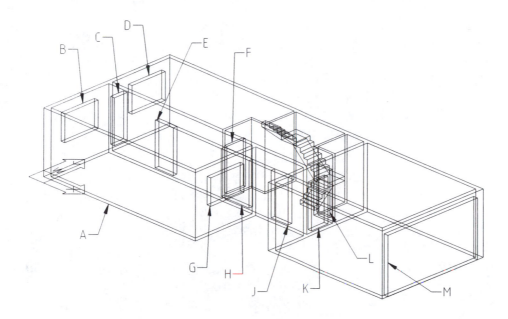

Figure 7.35 Window openings for the first level created

You have created 11 solid boxes to use as openings in the walls. At this stage, they are separate entities. To cut the openings, run the SUBTRACT command to subtract the boxes from the wall. You will not see any significant visual change in the model after subtraction.

[**Modify**]   [**Subtract**]

Command: **SUBTRACT**
Select solids and regions to subtract from...
Select objects: **[Select A (Figure 7.35).]**
Select objects: **[Enter]**
Select solids and regions to subtract...
Select objects: **[Select B, C, D, E, F, G, H, J, K, L, and M (Figure 7.35).]**
Select objects: **[Enter]**

After cutting the openings, turn on layer F1. See Figure 7.36.

<Data>   <Layers...>

Command: **DDLMODES**

Layer	
F1	On
F2	Off
ROOF	Off
STAIR	On
W1	On
W2	Off

Current layer: **W1**

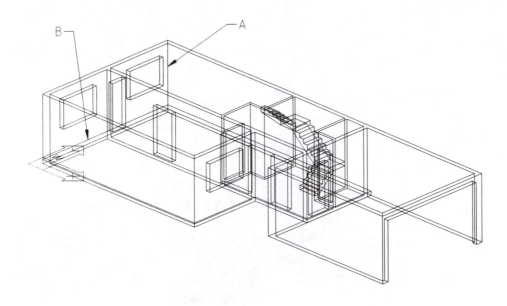

Figure 7.36 Layer F1 turned on

Run the UNION command to unite the wall and the floor. See Figure 7.37. The models of the first level and the staircase are complete.

[Modify]   **[Union]**

Command: **UNION**
Select objects: **[Select A and B (Figure 7.36).]**
Select objects: **[Enter]**

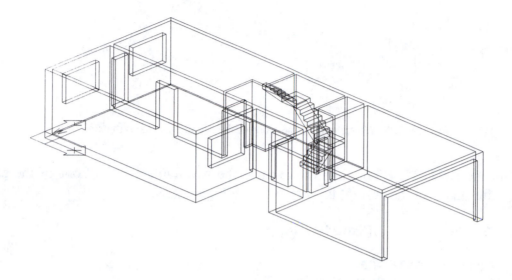

Figure 7.37 Completed first level and the staircase

7.7   3D Model for the Second Level

You will create the model for the floor, and then the model for the walls. After that, you will cut openings for the windows and doors.

Set the current layer to F2 and turn off all other layers except layer STAIR. See Figure 7.38.

<Data>   <Layers...>

Command: **DDLMODES**

Layer	
F1	Off
F2	On
ROOF	Off
STAIR	On
W1	Off
W2	Off

Current layer: **F2**

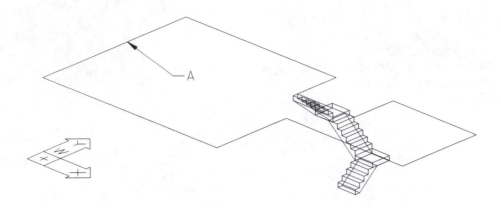

Figure 7.38 Wireframe for the floor of the second level and the staircase

Run the EXTRUDE command to extrude the wireframe for the floor of the second level. After that, set the current layer to W2. See Figure 7.39.

[Solids]　　　　　　**[Extrude]**

Command: **EXTRUDE**
Select objects: **[Select A (Figure 7.38).]**
Select objects: **[Enter]**
Path/<Height of Extrusion>: **-180**
Extrusion taper angle <0>: **[Enter]**

<Data>　　　　　　**<Layers...>**

Command: **DDLMODES**

Layer	
F1	Off
F2	On
ROOF	Off
STAIR	On
W1	Off
W2	On

Current layer: **W2**

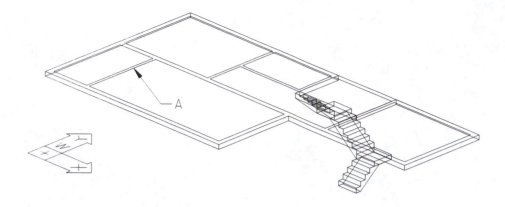

Figure 7.39 Floor for the second level created

Layer W2 contains the wireframe for the walls of the second floor. Use the EXTRUDE command to extrude this wireframe. After extrusion, use the MOVE command to translate the extruded solid a distance of 180 units in the negative Z direction. See Figure 7.40.

[Solids]   **[Extrude]**

Command: **EXTRUDE**
Select objects: **[Select A (Figure 7.39).]**
Select objects: **[Enter]**
Path/<Height of Extrusion>: **2880**
Extrusion taper angle <0>: **[Enter]**

[Modify]   **[Move]**

Command: **MOVE**
Select objects: **LAST**
Select objects: **[Enter]**
Base point or displacement: **0,0,-180**
Second point of displacement: **[Enter]**

The walls and floor for the second level of the house are complete.

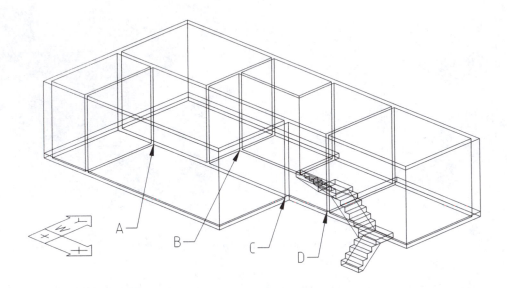

Figure 7.40 Walls of the second level created and translated

Next, you will cut window and door openings on the walls. Figure 7.41 is a dimensioned layout of the windows and doors. The lower edge of the windows are 1000 units from the floor surface, and the windows are 1200 units high. The height of the doors are 2000 units.

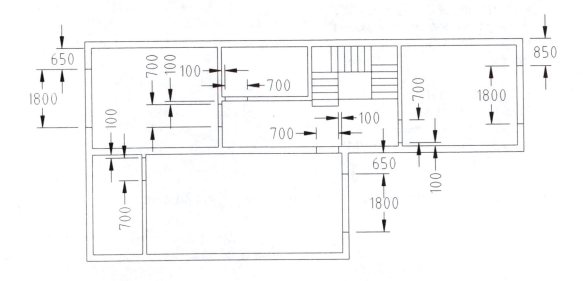

Figure 7.41 Dimensioned layout of the window and door openings

Using the dimensions shown in Figure 7.41, run the BOX command to prepare five solid boxes to use as door openings. See Figure 7.42.

[Solids]   **[Box]**   **[Corner]**

Command: **BOX**
Center/<Corner of box>: **FROM**
Base point: **END** of **[Select A (Figure 7.40).]**
<Offset>: **@0,-100,180**
Cube/Length/<other corner>: **@-120,-700,2000**

[Solids]   **[Box]**   **[Corner]**

Command: **BOX**
Center/<Corner of box>: **FROM**
Base point: **END** of **[Select B (Figure 7.40).]**
<Offset>: **@0,-100,180**
Cube/Length/<other corner>: **@-120,-700,2000**

[Solids]   **[Box]**   **[Corner]**

Command: **BOX**
Center/<Corner of box>: **FROM**
Base point: **END** of **[Select B (Figure 7.40).]**
<Offset>: **@100,0,180**
Cube/Length/<other corner>: **@700,120,2000**

[Solids]   **[Box]**   **[Corner]**

Command: **BOX**
Center/<Corner of box>: **FROM**
Base point: **END** of **[Select C (Figure 7.40).]**
<Offset>: **@-100,0,180**
Cube/Length/<other corner>: **@-700,200,2000**

[Solids]   **[Box]**   **[Corner]**

Command: **BOX**
Center/<Corner of box>: **FROM**
Base point: **END** of **[Select D (Figure 7.40).]**
<Offset>: **@0,100,180**
Cube/Length/<other corner>: **@120,700,2000**

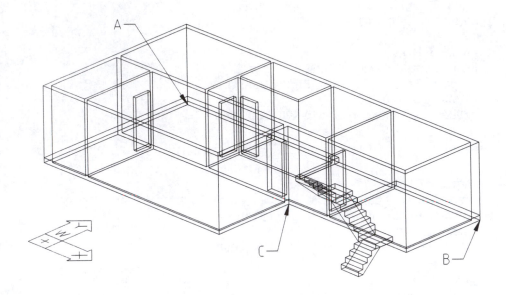

Figure 7.42 Boxes for the door openings created

Using the dimensions shown in Figure 7.41, create three solid boxes to use as window openings by running the BOX command. See Figure 7.43.

[Solids]   **[Box]**   **[Corner]**

Command: **BOX**
Center/<Corner of box>: **FROM**
Base point: **END** of **[Select A (Figure 7.42).]**
<Offset>: **@0,-650,1180**
Cube/Length/<other corner>: **@-200,-1800,1200**

[Solids]   **[Box]**   **[Corner]**

Command: **BOX**
Center/<Corner of box>: **FROM**
Base point: **END** of **[Select B (Figure 7.42).]**
<Offset>: **@0,-850,1180**
Cube/Length/<other corner>: **@-200,-1800,1200**

[Solids]   **[Box]**   **[Corner]**

Command: **BOX**
Center/<Corner of box>: **FROM**
Base point: **END** of **[Select C (Figure 7.42).]**
<Offset>: **@0,-650,1180**
Cube/Length/<other corner>: **@-200,-1800,1200**

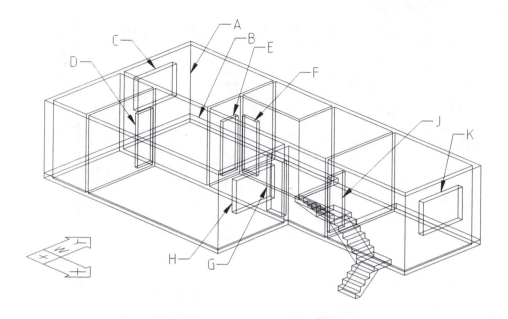

Figure 7.43 Boxes for the window openings created

Run the UNION command to unite the walls and the floor of the second level. Then, cut the window and door openings from the wall by using the SUBTRACT command. See Figure 7.44.

[**Modify**]   [**Union**]

Command: **UNION**
Select objects: **[Select A and B (Figure 7.43).]**
Select objects: **[Enter]**

[**Modify**]   [**Subtract**]

Command: **SUBTRACT**
Select solids and regions to subtract from...
Select objects: **[Select A (Figure 7.43).]**
Select objects: **[Enter]**
Select solids and regions to subtract...
Select objects: **[Select C, D, E, F, G, H, J, and K (Figure 7.43).]**
Select objects: **[Enter]**

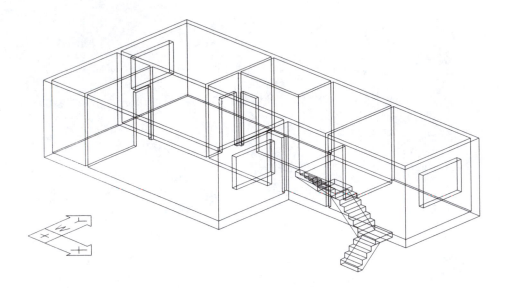

Figure 7.44 Completed second level

The second level of the house is complete.

7.8   3D Model for the Roofs

The final stage in creating the model is to work on the roofs, which have seven wireframes in three groups. The roof for the second level needs two groups of wireframes, and the roof for the first level needs one group of wireframes.

You will extrude all seven wireframes to solids of extrusion. After extrusion, you will create three solids of intersection. Finally, you will unite two solids of intersection to become the roof of the second level, and leave the third solid of intersection as the roof of the first level.

Set the current layer to ROOF, and turn off all other layers. See Figure 7.45.

<Data>   <Layers...>

Command: **DDLMODES**

Layer	
F1	Off
F2	Off
ROOF	On
STAIR	Off
W1	Off
W2	Off

Current layer: **ROOF**

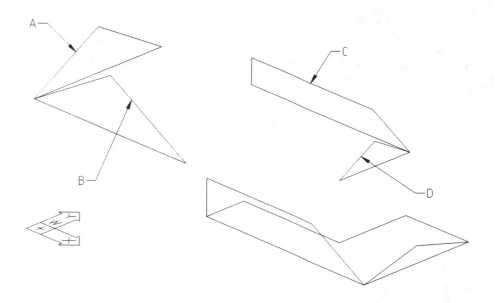

Figure 7.45 Wireframes for the roofs

Run the EXTRUDE command to extrude the wireframes for the roof of the second level. As mentioned before, the direction of extrusion depends on the Z direction of the UCS of the wireframes themselves, not the current UCS. See Figure 7.46.

[**Solids**]   [**Extrude**]

Command: **EXTRUDE**
Select objects: **[Select A (Figure 7.45).]**
Select objects: **[Enter]**
Path/<Height of Extrusion>: **7200**
Extrusion taper angle <0>: **[Enter]**

[**Solids**]   [**Extrude**]

Command: **EXTRUDE**
Select objects: **[Select B (Figure 7.45).]**
Select objects: **[Enter]**
Path/<Height of Extrusion>: **8600**
Extrusion taper angle <0>: **[Enter]**

[**Solids**]   [**Extrude**]

Command: **EXTRUDE**
Select objects: **[Select C (Figure 7.45).]**
Select objects: **[Enter]**
Path/<Height of Extrusion>: **3900**
Extrusion taper angle <0>: **[Enter]**

[**Solids**]   [**Extrude**]

Command: **EXTRUDE**
Select objects: **[Select D (Figure 7.45).]**
Select objects: **[Enter]**
Path/<Height of Extrusion>: **-9000**
Extrusion taper angle <0>: **[Enter]**

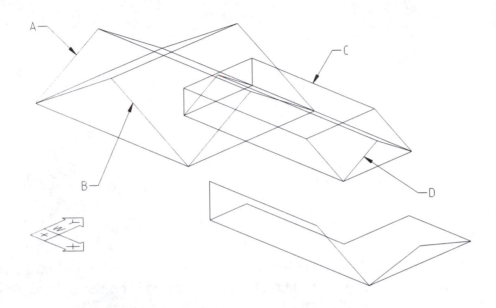

Figure 7.46 Four extruded solids created

You have created four solids of extrusion. Run the INTERSECT command to obtain two solids of intersection. See Figure 7.47.

[Modify]   **[Intersection]**

Command: **INTERSECT**
Select objects: **[Select A and B (Figure 7.46).]**
Select objects: **[Enter]**

[Modify]   **[Intersection]**

Command: **INTERSECT**
Select objects: **[Select C and D (Figure 7.46).]**
Select objects: **[Enter]**

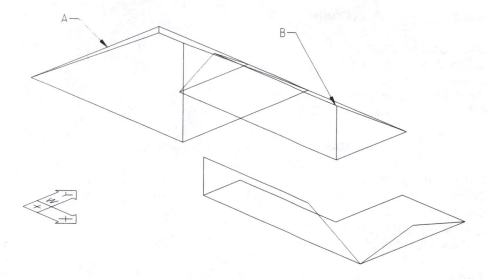

Figure 7.47 Two solids of intersections created

The two solids of intersection are components of the roof for the second level. Run the UNION command to unite them. See Figure 7.48.

[Modify]   **[Union]**

Command: **UNION**
Select objects: **[Select A and B (Figure 7.47).]**
Select objects: **[Enter]**

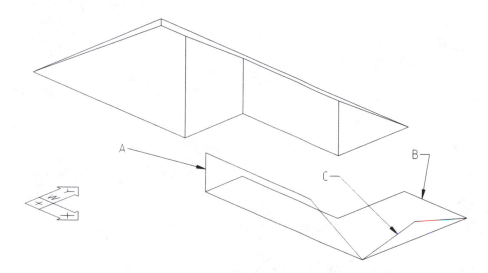

Figure 7.48 Roof for the second level completed

There are three wireframes left. Run the EXTRUDE command on them to form three solids of extrusion. See Figure 7.49.

[**Solids**]   [**Extrude**]

Command: **EXTRUDE**
Select objects: [**Select A (Figure 7.48).**]
Select objects: [**Enter**]
Path/<Height of Extrusion>: **5800**
Extrusion taper angle <0>: [**Enter**]

[**Solids**]   [**Extrude**]

Command: **EXTRUDE**
Select objects: [**Select B (Figure 7.48).**]
Select objects: [**Enter**]
Path/<Height of Extrusion>: **1700**
Extrusion taper angle <0>: [**Enter**]

[**Solids**]   [**Extrude**]

Command: **EXTRUDE**
Select objects: [**Select C (Figure 7.48).**]
Select objects: [**Enter**]
Path/<Height of Extrusion>: **-9000**
Extrusion taper angle <0>: [**Enter**]

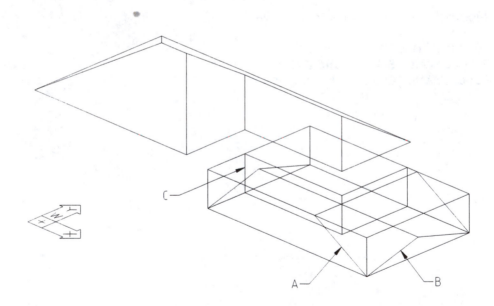

Figure 7.49 Wireframes for the roof of the first level extruded

Complete the roof for the first level by using the INTERSECT command on the three solids of extrusion. See Figure 7.50.

[**Modify**]   [**Intersection**]

Command: **INTERSECT**
Select objects: **[Select A, B, and C (Figure 7.49).]**
Select objects: **[Enter]**

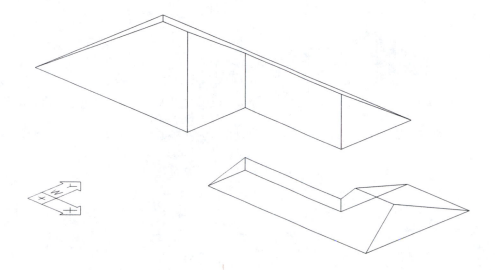

Figure 7.50 Completed roofs

The roofs for both levels are complete.

7.9   Completed Model

To view the entire model of the house, turn on all the layers and set current layer to 0. See Figure 7.51.

<Data>   **<Layers...>**

Command: **DDLMODES**

Layer	
F1	On
F2	On
ROOF	On
STAIR	On
W1	On
W2	On

Current layer: **0**

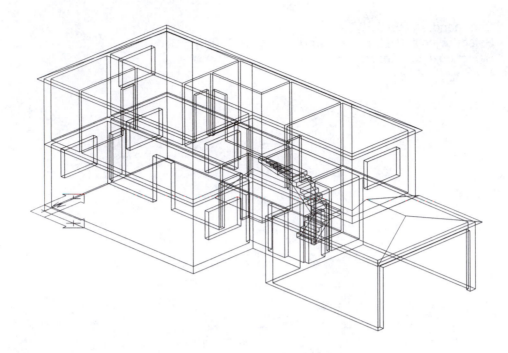

Figure 7.51 All layers turned on

To add details to the model, you will create four new drawings for the windows, doors, and gate to insert into the main drawing.

Save the current drawing, and start a new drawing by using the NEW command.

<File>   <New...>

Command: **NEW**

Using the dimensions shown in Figure 7.52, run the BOX command and the SUBTRACT command to create the window frame and the window panel.

After you have completed the drawing, use the BASE command to set the insertion base point.

Command: **BASE**
Base point: **[Select A (Figure 7.52).]**

Use the SAVE command to save the drawing under the name WINDOW.

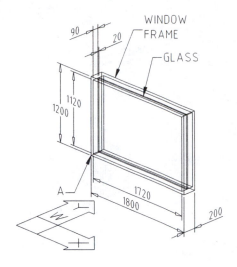

Figure 7.52 Window drawing

Similar to the window, start another new drawing. Using the dimensions shown in Figure 7.53, create a drawing called DOOR700. Remember to set the insertion base point to A (Figure 7.53) using the BASE command.

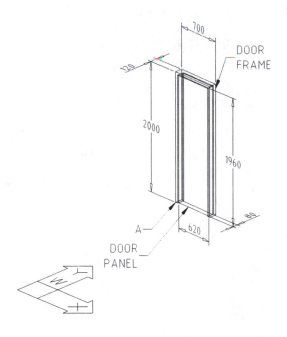

Figure 7.53 Door drawing

As shown in Figure 7.54, create another new drawing called DOOR900. Do not forget to set the insertion base point to A (Figure 7.54).

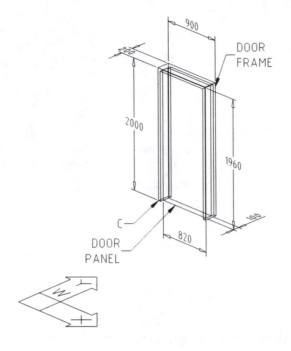

Figure 7.54 Second door drawing

Using Figure 7.55, make a drawing called GATE. The base point is at A (Figure 7.55).

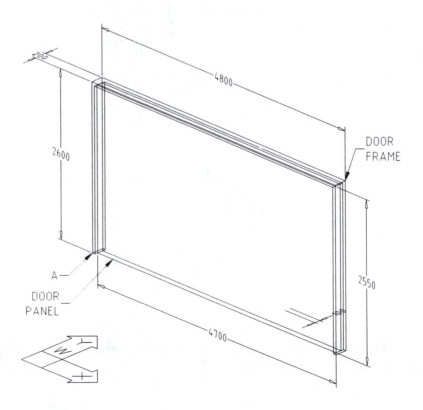

Figure 7.55 Gate drawing

After you make four drawings, open the drawing of the house again.

<File>   **<Open...>**

Set the current layer to W1 and turn off all other layers. See Figure 7.56.

<Data>   **<Layers...>**

Command: **DDLMODES**

Layer	
F1	Off
F2	Off
ROOF	Off
STAIR	Off
W1	On
W2	Off

Current layer: **W1**

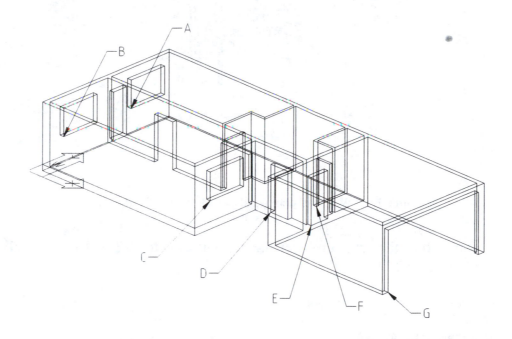

Figure 7.56 First level of the house

Run the INSERT command to insert the drawing WINDOW.

Command: **INSERT**
Block name (or ?): **WINDOW**
Insertion point: **END** of **[Select A (Figure 7.56).]**
X scale factor <1> / Corner / XYZ: **1**
Y scale factor (default=X): **1**
Rotation angle <0>: **90**

Repeat the INSERT command to insert the drawings, DOOR700, DOOR900, and GATE to this drawing. See Figure 7.57.

Command: **INSERT**

Block name	Insertion point	X scale	Y scale	Rotation angle
WINDOW	**B**	**1**	**1**	**90**
WINDOW	**C**	**1**	**1**	**90**
DOOR900	**D**	**1**	**1**	**0**
DOOR900	**E**	**1**	**1**	**90**
DOOR700	**F**	**1**	**1**	**0**
GATE	**G**	**1**	**1**	**90**

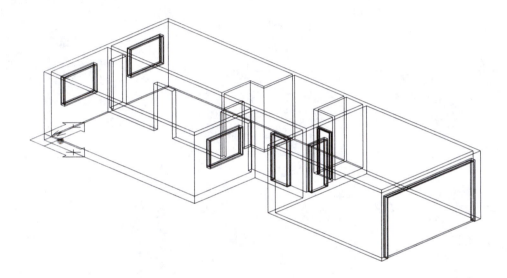

Figure 7.57 Windows, doors, and gate inserted to the first level

After completing the first level, set the current layer to W2 and turn off all other layers. See Figure 7.58.

<Data>   <Layers...>

Command: **DDLMODES**

Layer	
F1	**Off**
F2	**Off**
ROOF	**Off**
STAIR	**Off**
W1	**Off**
W2	**On**

Current layer: **W2**

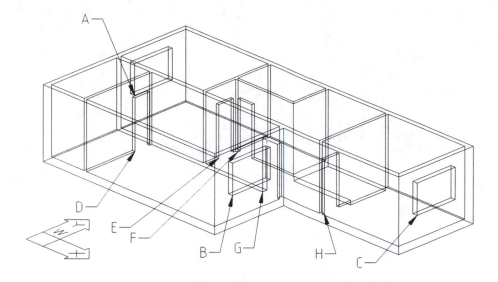

Figure 7.58   Second level of the house

The treatment of the second level is similar to the first level. Use the INSERT command to insert the drawings WINDOW and DOOR700. See Figure 7.59.

Command: **INSERT**

Block name	Insertion point	X scale	Y scale	Rotation angle
WINDOW	A	1	1	90
WINDOW	B	1	1	90
WINDOW	C	1	1	90
DOOR700	D	1	1	90
DOOR700	E	1	1	90
DOOR700	F	1	1	0
DOOR700	G	1	1	0
DOOR700	H	1	1	90

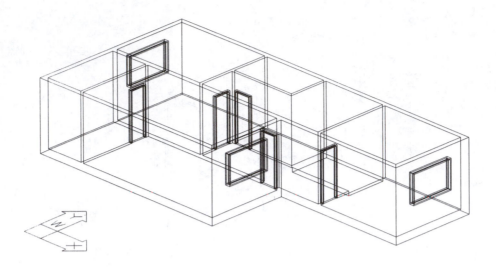

Figure 7.59 Windows and doors inserted to the second level

The model of the house is complete. Turn on all the layers. See Figure 7.60.

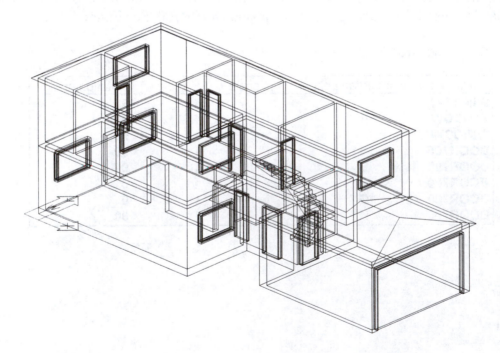

Figure 7.60 Completed house

7.10 Summary

In this chapter, you applied the constructive solid geometry commands with 2D entity creation and editing commands of AutoCAD Release 13 to create the model of a house.

By now, you should know how to create 2D layouts with multilines, how to convert multilines into regions for extrusion, how to cut openings in the extruded walls of the model house, and how to apply the constructive solid geometry techniques to architectural projects.

7.11 Exercise

To enhance your knowledge of using the solid modeling commands in architectural projects, you will build the solid model of another house. See Figure 7.61.

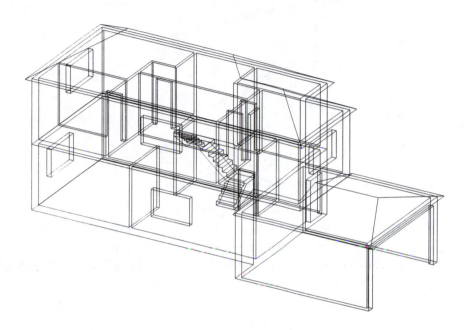

Figure 7.61 Completed house

Start a new drawing. Set up six additional layers:

Layer	Purpose
F1	Floor of the first level
F2	Floor of the second level
Roof	Roof
Stair	Staircase
W1	Wall of the first level
W2	Wall of the second level

All together, there are six groups of wireframes. Figure 7.62 shows all the wireframes required.

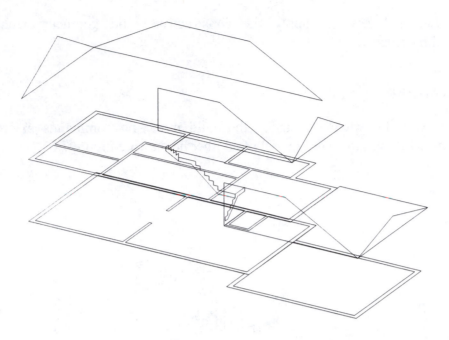

Figure 7.62 Wireframes required

Figure 7.63 shows the dimensioned layout of the walls of the first floor. Set layer W1 as the current layer. As shown in the drawing, create the entities as multilines. Then, explode them and convert them to a complex region for subsequent extrusion. The distance of extrusion is 2880 units.

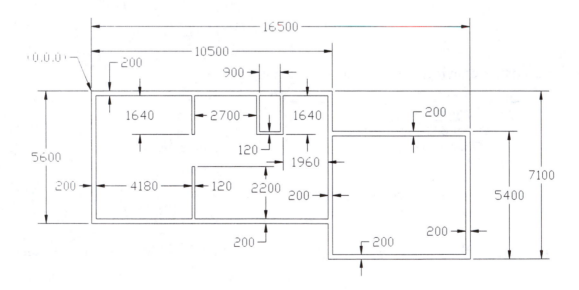

Figure 7.63 Walls of the first level

Set the current layer to F1. Then, create the wireframe for the floor of the first level as shown in Figure 7.64. The distance of extrusion for the floor is 180 units.

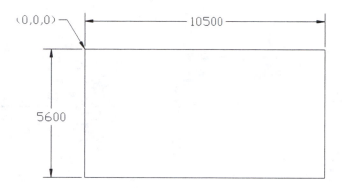

Figure 7.64 Floor of the first level

Place the UCS origin at (*0,0,2880). Set the current layer to W2. Then, create a complex region as shown in Figure 7.65. This is the walls for the second level. The distance of extrusion is 2880 units.

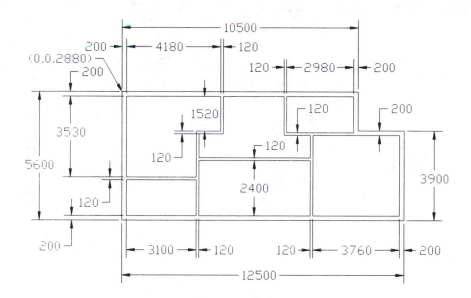

Figure 7.65 Walls of the second level

Keep the UCS origin at (*0,0,2880), and set the current layer to F2. As shown in Figure 7.66, create a region for the floor of the second level. The distance of extrusion is 180 units.

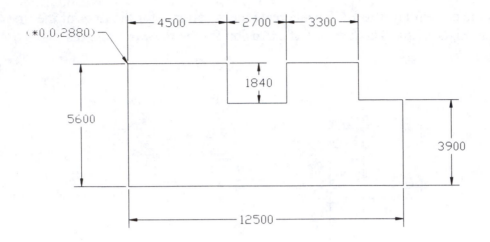

Figure 7.66 Floor of the second level

Set the current layer to STAIR. Then, create three sets of wires for making the solid model of the staircase. The dimensions of the staircase are identical to those of the house you have created in this chapter. Therefore, you can use the WBLOCK command to export the wireframe of that house to a file, and then use the INSERT command to import the wireframe to this house.

For your reference, Figure 7.67 shows the dimensions of the staircase. The distance of extrusion for each section of the staircase is 825 units.

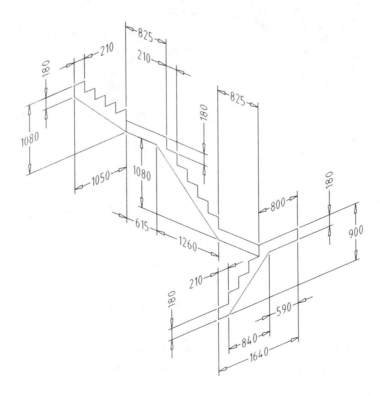

Figure 7.67 Staircase dimensions

Whether you prefer to insert the wires or to create the wires, you need to know the position of the staircase in relation to the other objects. Figure 7.68 indicates the relationship between the wireframes of the staircase and the wireframes of the walls of the first floor.

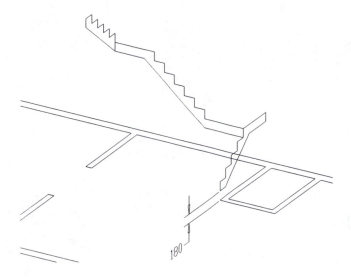

Figure 7.68 Position of the staircase

After making the wires for the staircase, set the current layer to ROOF. Then, create seven sets of wireframes for the roofs. Similar to making the staircase, you have to locate the UCS for each section of the roof. See Figure 7.69.

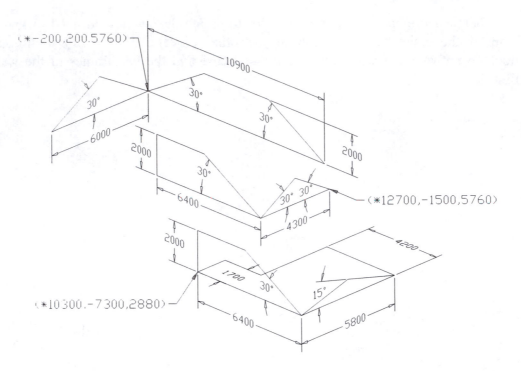

Figure 7.69 Roofs

The wireframes are complete. You can start to build the solid model for the house. Turn off layers W2, F2, and ROOF. Set the current layer to F1, and extrude the wireframes on layer F1 a distance of 180 units. Then, set the current layer to W1 and extrude the wireframes on layer W1 a distance of 2880 units. Finally, set the current layer to STAIR and extrude the three staircase sections a distance of 825 units.

After completing the staircase and the walls and floor of the first level, cut a number of openings as shown in Figure 7.70.

The height of the doors and gate are 2000 units and 2600 units, respectively. The windows are 1000 units from the top of the floor and are 1200 units high.

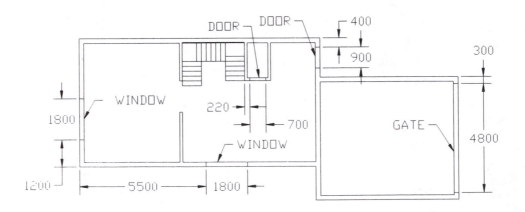

Figure 7.70 Window, door, and gate openings for the first level

Figure 7.71 shows the completed first level and the staircase.

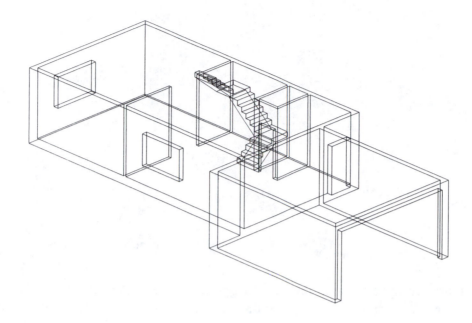

Figure 7.71 Completed first level and the staircase

The first level and the staircase are complete. Turn off the layers W1, F1, and STAIR. Set the current layer to W2 and extrude the entities on this layer a distance of 2880 units. Set the current layer to F2 and extrude the floor wireframes a distance of 180 units.

After making the floor and walls of the second floor, cut a number of door and window openings as shown in Figure 7.72.

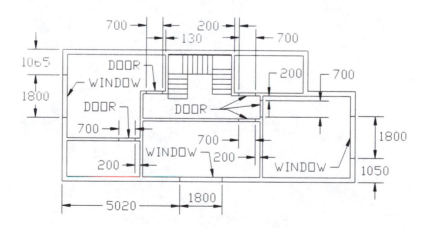

Figure 7.72 Window and door openings for the second level

Figure 7.73 shows the completed second level of the house.

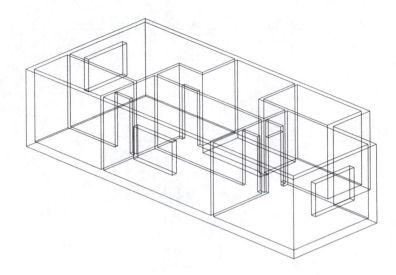

Figure 7.73 Completed second level

After completing the second level of the house, set the current layer to ROOF and turn off the layers W2 and F2. Extrude the seven wires. Then, form three solids of intersection, and unite two of them to form the roof of the second level. See Figure 7.74.

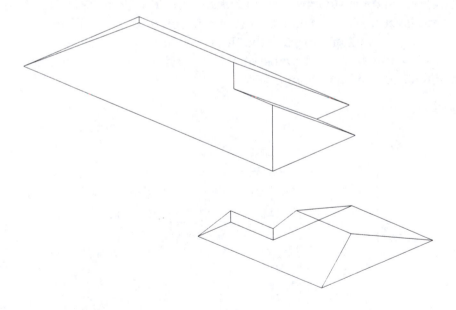

Figure 7.74 Completed roofs

The house is complete. Turn on all the layers. See Figure 7.61. Save your drawing.
To further refine the drawing of this house, you can insert the window, doors and gate that you prepared earlier in this chapter.

Appendix
Quick Reference

In the previous chapters, you practiced creating 3D solid models, and learned what you can do with a solid model. By now, you should be able to use the solid modeling commands of AutoCAD Release 13 as a tool to create 3D solid models to meet your design needs.

This appendix provides you with a summary of the AutoCAD Release 13 solid modeling commands for a quick reference.

A.1 Primitive Solid Creation Commands

REGION	creates regions from lines, arcs, circles, polylines, and ellipses.
BOX	creates a solid box.
CONE	creates a solid cone with a circular or elliptical base.
CYLINDER	creates a solid cylinder with a circular or elliptical base.
SPHERE	creates a solid sphere.
TORUS	creates a solid torus.
WEDGE	creates a solid wedge.
EXTRUDE	creates a solid by extruding a polyline or region.
REVOLVE	creates a solid by revolving a polyline or region about an axis.

A.2 Boolean Operation Commands

UNION	creates a complex solid from two or more solids by uniting the solids, or creates a complex region from two or more regions by uniting the regions.
SUBTRACT	creates a complex solid from two or more solids by subtracting one or more solids from a solid, or creates a complex region from two or more regions by subtracting one or more regions from a region.
INTERSECT	creates a complex solid from two or more solids by yielding their intersecting portion, or creates a complex region from two or more regions by yielding their intersecting portion.
INTERFERE	locates the interference of two or more solids, and creates a solid from the common volume of the selected solids.
SLICE	slices a solid into two solids.

A.3   Solid Editing Commands

CHAMFER	creates a beveled edge at the edge of a solid.
FILLET	creates a rounded edge at the edge of a solid.
ALIGN	translates an object three-dimensionally to align with another object.
3DARRAY	creates a 3D array of objects.
MIRROR3D	creates a mirror image of objects about a 3D plane.
ROTATE3D	rotates objects about a 3D axis.

A.4   Documentation Commands

SOLVIEW	creates floating viewports and layers to generate 2D drawing views.
SOLDRAW	generates 2D drawing views in floating viewports that are created by the SOLVIEW command.
MVIEW	creates floating viewports.
SOLPROF	generates 2D drawing views in floating viewports that are created by the MVIEW command.

A.5   Solid Utility Commands

MASSPROP	evaluates the mass properties of a solid or region.
SECTION	creates a region across a specified plane on a solid.
AMECONVERT	converts an AME solid objects to an AutoCAD native solid.
3DSOUT	exports to a 3D Studio file format.
STLOUT	exports a solid to STL (Stereolithography) file format.

Index